Microscopy of Materials

Microscopy of Materials

Modern Imaging Methods Using Electron, X-ray and Ion Beams

D. K. Bowen
Department of Engineering

C. R. Hall
Department of Physics,
University of Warwick

M

© D. K. Bowen and C. R. Hall 1975

First published 1975 by
THE MACMILLAN PRESS LTD
London and Basingstoke
Associated companies in New York Dublin
Melbourne Johannesburg and Madras

SBN 333 15495 9   (hard cover)
333 18703 2 (paper cover)

Printed and bound in Great Britain by
A. WHEATON & CO.
Exeter

Contents

Preface

Microscopy is a craft that requires an 'on-line brain'. This is why it is one of the most satisfying activities available in the experimental sciences. Because it is a craft, it demands a high level of skill in preparing the specimens and in adjusting the instruments to extract the fullest information; but more important is the knowledge that the experimenter is continually using: his familiarity with the principles and practice of the instrument itself so that its control becomes second nature and his understanding of the theory of image formation in the microscope, which enables him to select the conditions appropriate to the problem—and to be confident of his results.

But however fascinating the practice, the *raison d'être* of microscopy is that it is a tool for investigating the microstructure of materials. The establishment over many decades of some of the relationships between the microstructure and the macroscopic properties of materials has had profound consequences in scientific understanding and in industrial practice. Increased understanding of traditional materials has led to their safer, better or cheaper use; new types of material have been developed with spectacular mechanical or electrical properties, and better knowledge has meant better control of the production of both materials and components. During the period of this advance, both energy and raw materials were relatively abundant and cheap; this era has ended and the efficient use of such resources is now imperative.

A formidable array of techniques of microscopy is now available to the investigator. Unfortunately a technical and theoretical description of these methods is often only to be found in very advanced forms, such as original papers or monographs aimed at the experienced professional scientist. Our aim in this book has been to write a balanced account of all the important advanced methods of microscopy, beginning at a level convenient for undergraduates. In the main chapters we have covered scanning and transmission electron microscopy, the various microanalytical techniques, X-ray topography and field–ion microscopy. In the final chapter we have treated more briefly some rather rare methods, such as photoemission microscopy, and have endeavoured to survey the most recent developments, some of which may become important in the future. We have discussed the principles and construction of each instrument, have treated the theory of contrast from various types of specimen in some detail and have given examples of the application of each technique, chosen to cover a wide range of materials and types of problem. We hope that we have answered two rather different questions: what can this instrument be used for, and, which technique is best for this particular problem? We have not attempted to write a buyer's guide or operator's manual—although prospective buyers or operators should know most of the contents of this book—and so we have ignored such points as the idiosyncrasies of instruments made by particular manufacturers, purely technical problems such as the design of special stages, or matters of historical interest only such as the once severe problem of specimen contamination in the electron microscope (which has now been largely eliminated by better vacuum technology). SI units are used throughout, with the addition of the electron volt and the torr.

Although optical microscopy is an important method for the study of microstructure it will doubtless be abundantly familiar to almost every reader of this book and, moreover, it is already well covered in textbooks at a number of levels. We have therefore excluded optical microscopy from our treatment, along with various non-imaging methods of X-ray, electron or neutron diffraction. However, chapter 1 contains a brief review of the microstructural information available from these techniques and their advantages and limitations in order that the remaining chapters may be appreciated in a proper context.

The book as a whole is intended to be appropriate for a second- or third-year undergraduate course in materials science, metallurgy, physics or engineering science. We have assumed a basic knowledge of the structures and microstructures of materials, of common types of defects in crystalline materials and of elementary diffraction (essentially Bragg's law and the reciprocal lattice construction). The mathematical level required is generally no more than school-level calculus and algebra with the addition of some vector algebra and some elementary statistics and Fourier analysis; a knowledge of the last two topics is not central to comprehension of the book as a whole. The level of knowledge assumed in different chapters, and suggestions for preliminary reading for the 'non-standard reader' are discussed in section 1.3. Of course, a consistent level of treatment leads to an inconsistent level of rigorous interpretation simply because some phenomena are more complex than others. Thus, we have derived the basic equations of electron diffraction in chapter 5, but in chapter 6 the corresponding theory for X-ray diffraction has only been qualitatively explained by analogy with the electron case. We were unwilling to assume familiarity with vector calculus, Maxwell's electrodynamic field equations and some slightly tricky manipulation of Fourier series, and moreover, the importance of the subject to a book at this level would not justify such complexity. However, the treatment of both electron and X-ray diffraction is in terms of the dynamical theories, without which the comprehension of electron microscopy contrast becomes superficial and that of X-ray topography almost non-existent. We hope that we have succeeded in making these theories accessible to the undergraduate.

We have had two other classes of reader very much in mind. One is the graduate in some other scientific discipline who finds himself doing research or development work in a university, national laboratory or industrial laboratory and becomes involved in the study of the microstructure of materials. He should find here a reasonably rapid conversion course on the advanced observational techniques, that will enable him to select those applicable to his own field. The other is the graduate in materials science, metallurgy or a related field who finds himself (in, say, an industrial laboratory) expected to work on or even take charge of advanced equipment for microscopy with which he had only a passing acquaintance as a student. The equipment described in this book is expensive; the cheapest is probably one of the simple X-ray cameras which with its X-ray unit would cost about two years' salary of the new graduate, while the dearest (the high-voltage electron microscope) might be purchased with his lifetime's salary if he were particularly successful. To avoid wasting the resources that these instruments represent they must be used intelligently, and we must emphasise that the best and most meaningful results are obtained from modern microscopes only when the experimenter thoroughly understands his problem, his instrument and the theory of its operation.

It is a pleasure to express our gratitude to all those who have helped us in various ways in the preparation of this book. First, we should like to thank those who taught us and stimulated our own interest in the science and art of microscopy, in particular Professors P. B. Hirsch, F. R. S. and J. W. Christian, F. R. S. and Drs M. J. Whelan, A. Howie and L. M. Brown. Next, we are happy to acknowledge discussions with many colleagues on various aspects of the techniques described in this book and would especially mention Drs G. R. Booker, C. van Essen, J. Miltat and G. Slodzian. We are most grateful to those who have provided us with photographs or diagrams illustrating various aspects of their work; this has meant that we have been able to draw on a much greater variety of examples than would otherwise have been possible. Specific acknowledgements appear in the figure captions for these illustrations. We should like to thank Mr C. G. Lovatt for valuable assistance with the preparation of the photographs and also the many Warwick University students of materials science, physics and engineering science on whom much of the contents of this book has been tested. Finally, we now understand very clearly why authors invariably thank their wives for their assistance and forbearance. We likewise thank them.

University of Warwick, Coventry
January 1975

D. K. BOWEN
C. R. HALL

The Study of Microstructure

Materials science is totally concerned with the structure of materials, with the understanding and control of structure and with the prediction and understanding of its ultimate consequences—the macroscopic properties of matter. Materials scientists, therefore, use extensively those techniques that have been developed to investigate the structure of matter at all levels from the macroscopic down to the atomic. It is commonly found that most of the significant features of the structure of a material are of a 'microscopic' size, that is, from a few tens of microns down to atomic dimensions. Techniques for investigating structure at this level may be classified into methods based upon the analysis of diffraction patterns, which give information averaged from a relatively large sample, and imaging methods, which give detailed information about a small region. For very many applications the imaging methods are more useful, since the local arrangement of the microstructure is often extremely important. A few examples will make this clear. Plastic flow in a crystalline solid is controlled by the generation and movement of dislocations and their interaction with other microstructural features. A solid-state precipitate may greatly improve the mechanical properties of a material if it is finely dispersed within the grains, but can cause catastrophic brittleness if it collects at a grain boundary. The remarkable electrical properties of a complex integrated circuit depend very much upon small variations of composition within its semiconductor material; in contrast, the device may fail if the distribution of components is accidentally altered by preferential diffusion along crystal defects. The economic feasibility of a superconductor in an alternating-current application may depend upon the ability of groups of lattice defects and precipitates in the solid to 'pin' the magnetic flux lines and minimise the hysteresis losses. Finally, the plague of corrosion, costing not a few lives and vast sums of money every year is caused by reactions at small, localised sites on the surface of a material.

This book is concerned with the techniques for obtaining images of lattice defects, phase distributions, electrical and magnetic phenomena, surface structures and many other aspects of the microstructure. In particular, we discuss the more modern and advanced methods that are the more difficult to approach. An overall view of these techniques is given in section 1.3 so that the context of individual chapters in this book may be more readily appreciated. First, however, it is useful to review the other microstructural methods such as the traditional technique of optical microscopy together with the methods based upon the analysis of diffraction patterns, in order to bring out the advantages and limitations of the newer methods. Although optical microscopy and the X-ray diffraction methods have, on the whole, been established for longer than the modern imaging techniques, they are in no way obsolete. Advances in instrumentation are still occurring; for example, the modern optical microscope or X-ray diffractometer is much faster, more reliable and more convenient to use than the models of ten or twenty years ago, even though the basic methods and ultimate accuracies have not greatly changed in this period. These experimental methods are useful, not only because they complement the more modern equipment, but also because they remain as the main equipment for much research. They will therefore be summarised in the following sections, and assessed from the point of view of their ability to yield microstructural information (remembering that they have other uses as well) although the techniques themselves will not be described in any detail.

1.1 OPTICAL MICROSCOPY

The optical microscope is of course an imaging instrument, and it is the primary tool in many materials investigations. Its characteristics may be briefly summarised: it has a resolution of about 250 nm, a depth of field of the same order and is relatively cheap and easy to operate. The interpretation of the images is usually straightforward although, as with the more advanced methods, a careful study of the optical principles will help to avoid pitfalls and increase the utility of the technique. A considerable number of mechanisms is available for creating contrast in the image; these are based upon the reflection, diffraction, interference and polarisation of light waves, and the effects may be enhanced by various surface treatments such as selective etching, staining or coating.

Optical microscopy is particularly good in comparison with more recent techniques in two types of situation. The first is fairly obvious and occurs when the specimen causes effects specifically limited to the visible waveband. Examples are the rotation of the plane of polarised light by the symmetry or by the magnetisation of a crystal; the creation of differently coloured regions in a material by a phase transformation or a particular type of structural defect; or the phenomenon of optical fluorescence. The second field in which it works well is in the measurement of surface relief. Interferometric methods in the optical microscope are extremely good. It is easy to see steps as small as 5 nm by interference contrast microscopy and actual measurements have been made by multiple-beam interference of steps only 0.3 nm high. None of the methods described in this book can do so well on bulk specimens.

Optical microscopy has, of course, many limitations, the most serious being that its lateral resolution is rather poor (much worse than its vertical resolution using interferometry). It is therefore inadequate for studying the many effects occurring on a very fine scale in materials. Even when studying surface relief it may be preferable to use a scanning or transmission electron microscope (in the latter case, with a replica of the surface) and accept a poorer vertical resolution in order to obtain a far higher lateral resolution.

The next major limitation is that of transparency. Few materials are transparent to visible light and successive sectioning must be used in order to observe the three-dimensional microstructure of an opaque material, whereas the higher penetrative power of electrons and X-rays can be used to reveal such features directly. An alternative approach is preferentially to dissolve one constituent of the structure. This can show the full structure of the remaining constituents very well, but requires the use of a microscope with a much higher depth of field than that of the optical microscope.

The analytical information obtained from the optical microscope is only occasionally quantitative, and then only when the system under study is very well characterised. On the other hand, the methods described in chapter 3 are capable of reasonably accurate quantitative analysis on a microscopic scale even when nothing is initially known about the system. Nor can the optical microscope provide much information about crystallographic properties of the material or about crystal lattice defects. The orientation of a grain can sometimes be determined from the shape of etch pits, and the sites of surface outcrops of dislocations can also be found by etching, in some cases. In transparent materials, dislocations may be made visible inside a crystal, either by decorating the dislocations with impurity precipitates (historically, this was the first technique used to study dislocations—in silver chloride) or by using polarised light and observing the birefringent regions created by the stress field around the dislocation. However, the methods of electron microscopy and X-ray topography are far more general. They not only reveal individual dislocations and stacking faults with ease in a wide variety of materials, but also can be used for quantitative measurements such as the determination of Burgers vectors, fault vectors and dislocation densities.

Finally, perhaps the most important point is that many more contrast mechanisms become available when electron, X-ray and ion beams are used for microscopy. The modern techniques extend the type of information that is obtainable into wholly new fields.

1.2 DIFFRACTION TECHNIQUES

The non-imaging methods of diffraction, such as the Debye–Scherrer method for X-rays, do give a lot of structural information. The difference between the types of information given by imaging and non-imaging methods can be understood from figure 1.1, which shows the relation between the diffraction pattern and the image. When a specimen is

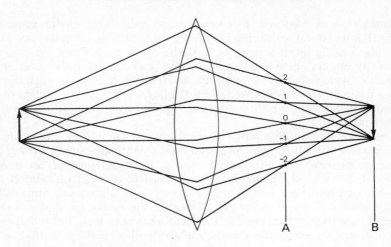

Figure 1.1. The relation between the image and the diffraction pattern. The various diffracted orders, labelled 0, 1, 2, leave the object as parallel bundles of rays. These are brought to a focus in the back focal plane A of the lens, and combine to form the image in plane B

illuminated with parallel, monochromatic radiation, the angle at which a beam is diffracted depends only on the wavelength of the radiation and the spacing and orientation of the microstructural features that give rise to the diffraction. The diffracted beams from similar structures in the specimen are therefore all parallel and are consequently brought to a focus in the focal plane of the lens. This is the plane A in figure 1.1, and the whole diffraction pattern is formed in this plane. If the rays are traced further, it is seen that an image results from the combination of rays from different diffracted beams. The image is formed in plane B. The information content of the diffraction pattern and the image in the microscope must be identical (unless an aperture is placed between A and B to restrict the information) but it is differently distributed. The diffraction pattern contains information that is averaged over the whole specimen whereas in the image the point-to-point distribution of this information is reconstituted. The relation between the diffraction pattern and the image can also be expressed mathematically, since one is effectively the Fourier transform of the other. These arguments are quite general; for example, if no lens is involved (as in the case of X-ray diffraction and X-ray topography) the system simply becomes equivalent to the pinhole camera. However, it is not possible merely to record the diffraction pattern or image and use the one to reconstitute the other, since at these enormously high frequencies there is no known method of recording the relative phase of the waves at different points of the diffraction pattern or of the image. Phase information is required before the Fourier transform can be carried out. In practice therefore, separate techniques are used for diffraction patterns and images and the information given in each is different.

With the help of these generalisations we can identify the spheres of interest of the diffraction and imaging techniques. Diffraction methods will be best when information averaged over the whole specimen is required—the outstanding example is the application of diffraction methods to the determination of crystal structure. Imaging methods are not very much use for this, since their resolution and image contrast are not usually quite high enough to resolve the unit cell itself (the method of 'real-space crystallography' mentioned in section 5.3.2 does obtain structural information from the electron microscope image, but it does so by means of the symmetry and intensity of certain diffraction fringe (extinction contour) intersections rather than by imaging the unit cell). High-resolution electron or field-emission microscopy can help to determine the overall shape of a complex biological molecule and so help the structural analysis to some extent, and the field–ion microscope does form an image of atoms but unfortunately can only handle a limited range of materials (whose structures are already well known). In general, any aspect of the structure that occurs throughout the specimen can affect the diffraction pattern, but imaging methods must be used to determine the detailed distribution of microstructural features.

1.2.1 X-ray Diffraction

The microstructural applications of X-ray diffraction are summarised in table 1.1. There are obviously considerable possibilities, and also some overlap between the diffraction and imaging methods. For example, one may measure particle sizes by small-angle scattering on the diffraction pattern or (with patience) by imaging and measuring a large number of the

Table 1.1. Microstructural applications of X-ray diffraction patterns (Techniques that do not give microstructural information are not included)

Microstructural application	Diffraction technique
Preferred orientation, texture determination	angular distribution of intensities of Bragg reflections
Lattice parameter measurement (hence solid solubility, vacancy concentration, etc.)	measurement of Bragg angles
Structure analysis	angles and intensities of Bragg reflections
Particle size in range 2–50 nm	(1) low-angle scattering (2) line broadening
Dislocation density, $\geqslant 10^4 \, mm^{-2}$	line broadening
Stacking-fault density	line broadening
Internal stress	line broadening
Solid solutions	diffuse scattering (between Bragg reflections)
Thermal vibrations	(thermal) diffuse scattering

particles, and it is better to regard the techniques as complementary rather than competitive. Very often it is found that both diffraction and imaging methods give more information when used together than either can alone. From table 1.1, it is evident that there is an embarrassingly large number of causes of X-ray line broadening; an electron microscope investigation could tell with some certainty the actual causes in a particular specimen, and then X-ray measurements could be made to obtain quantitative data. A similar procedure could be followed for the measurement of particle sizes where the sizes are not all the same, or where there is a distribution of particle shapes. Some of the different contributions to one effect (such as line broadening) can in fact be separated by mathematical deconvolution techniques based upon the different response of the various contributions as the Bragg angle changes. However, this process can become quite complicated and some of the techniques in table 1.1 have dropped out of favour—a trend which could perhaps be reversed if the possibility of combining them with the modern imaging methods were more generally realised. The advantage of using X-rays to make some of these measurements is that the results are averaged over far larger volumes of the sample than is possible with, say, the transmission electron microscope.

The methods based upon diffraction patterns break down when the feature under study is not present in sufficient concentration and also when the details of the distribution are important. One needs a dislocation density of at least $10^4 \, \text{mm}^{-2}$ ($10^6 \, \text{cm}^{-2}$) to detect any effect on the widths of lines in the diffraction pattern, but the imaging X-ray topographic method can show a single dislocation in an otherwise perfect crystal. Likewise the electron microscope can be used to decide whether a fine precipitate lies mainly within the grains or in the grain boundaries; although this information must be contained within the appropriate diffraction pattern, it is quite inaccessible.

1.2.2 Electron and Neutron Diffraction

These are clearly in the same class as X-ray diffraction experiments in that they must yield averaged information, but they extend the range of such experiments since X-rays, electrons and neutrons all interact differently with matter. Electron scattering factors are generally higher than those of X-rays, in particular for light elements; neutron scattering factors, while not showing so great a variation overall as either X-ray or electron scattering factors, are often very different for elements of adjacent atomic number (or indeed for different isotopes of one element), which will have rather similar X-ray or electron scattering powers. These types of radiation can, therefore, be used to sort out specific problems outside the range of the more ubiquitous X-ray techniques. Neutrons are also particularly useful in the study of magnetic materials since they themselves have a magnetic moment.

No imaging method exists for neutron radiation, nor is it likely that one will be invented. However, all electron microscopes can provide both images, and electron diffraction patterns from selected areas of these images. As will be seen in chapter 5, this is a very powerful combination; it emphasises the complementary nature of imaging and diffraction methods and suggests that much more could be achieved by combinations of techniques, even on separate instruments.

1.3 MODERN IMAGING METHODS

The techniques discussed in the main chapters of this book are briefly reviewed in this section in order to establish an overall view both of the subject and of the book. We shall also point out the links between the different chapters, which are not all self-contained (although cross-references are used to assist the reader who wishes to learn about one particular technique). Each chapter contains discussions of the principles of operation of the instruments, the important features of their design (in so far as this affects the performance or imaging properties), significant parameters such as the resolution, a discussion of the mechanism and theories of image contrast and a number of examples of the applications of the microscope. In some cases, we have deemed it appropriate to devote a separate section to the applications, in others the discussion of image contrast naturally brings out the applications. It is worth emphasising the importance of understanding image contrast since it is a topic that is too often taken for granted. Even in optical microscopy, an appreciation of the relatively simple theory undoubtedly leads to better use of the instrument. In the microscopes described in this book, the mechanisms of image contrast are rarely obvious and often very complicated, but if they are not understood the techniques are useless. For example, a dislocation can appear in an electron microscope as a black line—but so can bends in the specimen, changes in its thickness or dirt on its surface. The line can only be identified as a dislocation when the detailed differences between the various possible black lines (such as their widths, the small intensity variations along their lengths, their response to different illuminating conditions) are comprehended. Moreover, when the mechanisms of image contrast are understood the applications become more obvious; thus the knowledge that contrast can be obtained in the scanning electron microscope from electrical potentials on the surface of the specimen immediately suggests its use for the study of integrated circuits in operation. Contrast theory, therefore, forms an important part of each chapter, and it must be understood before the technique can be discussed or used with any confidence.

1.3.1 Scanning Electron Microscopy

In this instrument, commonly called the SEM, images are formed by the collection and amplification of electrons backscattered or emitted from the surface of a bulk specimen (or transmitted through a thin foil) when a high-energy electron beam is scanned across its surface. A magnified image is formed and the SEM can be used in the same sort of way as an optical microscope; however, its resolution is vastly superior (about 15 nm) as is its depth of field (up to several millimetres). The SEM would be worth using if it were only as a 'super magnifying glass', but it has many other powerful features. The complexity of the interaction of electrons with solids means that many other microstructural features give image contrast under the appropriate conditions, for example surface electric or magnetic fields, composition variations, crystal orientation, electron-fluorescent phases. The SEM is consequently widely used in research and industrial development or in trouble-shooting.

Chapter 2 also contains the basic electron optics used in chapters 3, 4, 5 and 8 such as the production of electron beams, the properties of magnetic lenses and the detection and amplification of electrons. Purely technological problems such as the design of vacuum systems, airlocks or high-tension supplies, are not treated in this book.

1.3.2 Microanalysis

Chapter 3 describes a group of instruments used to identify and measure the chemical composition of a solid. The volume analysed may be as small as 1 μm^3 and the analysis may be presented as an image of the distribution of a given element, so these instruments give truly microscopic analysis. The most important is the electron probe microanalyser; it is currently the only one that gives reliable quantitative as well as qualitative data. It is constructed similarly to the SEM and performs chemical analysis by measuring the wavelength and intensity of the characteristic X-radiation emitted when a solid specimen is struck by an energetic electron beam; the analysis can be made accurate to about ± 1 per cent for elements with atomic numbers greater than about 12. The secondary ion emission microscope, the Auger microscope and the laser microprobe, are more recent developments. The Auger instrument is again similar to the SEM (but must employ an ultra-high-vacuum system); it collects and counts the electrons emitted in a certain energy band that characterises the emitting atoms. In the ion microscope, an incident ion beam ejects ions from the specimen surface. This secondary ion beam is sorted in a mass spectrometer and focused to form a mass image. Both these newer methods offer advantages over the electron probe microanalyser, especially for the study of surfaces and the analysis of light elements, but they cannot yet provide so accurate a quantitative analysis. The laser microprobe uses a laser beam, located and focused by an optical microscope, to volatilise a few μm^3 of the specimen; metallic elements can then be identified by analysing the spectral emission in the visible band. This method is neither so sensitive nor so general as the others, but the apparatus is a lot cheaper and its accuracy is reasonably good (around 10 per cent).

This chapter relies upon the description of the SEM in chapter 2, in particular for the design of the electron microprobe since these two instruments have many similarities.

1.3.3 Transmission Electron Microscopy

The optics of this instrument are analogous to those of the transmission optical microscope except that electrons are used instead of visible light. Since electrons interact strongly with matter, specimen thicknesses are restricted to the range 100 nm–5 μm (the thickness within this range depends upon the material and the electron energy). The high electron energies used, 100 keV–1 MeV, mean that the wavelength is very short and the resolution consequently very good, about 100–200 pm, although individual atoms only show sufficient contrast to be imaged in rather special circumstances. The transmission electron microscope can reveal almost all the important microstructural features in crystalline or amorphous materials. In crystalline materials, much use is made of the diffraction of electrons, both by using the diffraction pattern to identify the structure or orientation of a feature (perhaps only 1 μm diameter) and by exploiting the diffraction process to give identifiable image contrast from features such as dislocations, stacking faults or small precipitates.

The construction and operation of the transmission electron microscope is treated in chapter 4, which again uses the electron optics of chapter 2. Chapter 5 deals with the theory of image contrast and the applications of the microscope, and some aspects of the theory are used in chapter 6.

1.3.4 X-ray Topography

Images are formed in X-ray topography (chapter 6) by recording the intensity of a diffracted beam over the whole surface of a crystal that has been uniformly illuminated by an incident beam. Both reflection and transmission methods are used but the latter are more versatile; they reveal the same microstructural features as the transmission electron microscope, but the scale is larger and the resolution poorer (about 1 μm, limited by several technological factors). X-ray topography is, therefore, particularly useful for the study of long-range phenomena such as lattice rotations during deformation or for investigating highly perfect crystals such as those used in the semiconductor or laser industries.

'X-ray microscopy' would really have been a better name for this group of techniques, but this term is generally used to mean high-resolution radiographic rather than diffraction methods. The theory of image contrast in X-ray topography is mathematically and conceptually quite difficult and is not treated in detail. Instead, the essential ideas needed to interpret the images are developed qualitatively, by arguments based on the similarity of the theory to the dynamical theory of electron diffraction, which is given in chapter 5.

1.3.5 Field–Ion Microscopy

Chapter 7 describes the instrument that has the highest useful resolution of all microscopes, since it can image individual atoms. Helium ions are usually the imaging medium; the magnification is essentially geometric, since the ions are accelerated in straight lines between the very small needle-tip specimen and the large fluorescent screen. The electric-field perturbations caused by atoms lying at the edges of atomic planes in the specimen result in these atoms being imaged.

Clearly, there are very severe constraints upon the type of specimen that can be used in the field–ion microscope, and the conditions inside the instrument (in particular the very strong electric field) are also highly artificial. Nevertheless, only by this microscope is it possible directly to study atomic configurations around features such as dislocations, stacking faults, grain boundaries, vacancies or vacancy clusters, or atomic movements during surface processes such as oxidation.

This chapter is essentially self-contained.

1.4 GUIDE TO PRELIMINARY READING

The techniques described in this book are used in many different fields and are relevant to courses in many different subjects. It is, therefore, appropriate to indicate texts that may be used for background reading on the topics relevant to the different chapters.

First, there is no point in studying advanced techniques for the investigation of microstructure without at least an elementary appreciation of the significance of the microstructure of materials. This is no problem for a materials science or metallurgy student, or a physics student who has taken an option in this field, but the graduate in another discipline (pure physics, electrical engineering, etc.) who is using this book as part of a 'conversion course' should first read a standard text on materials science or metallurgy, such as Van Vlack (1970), or Reed-Hill (1973), unless he is concerned only with the principles of design and construction of the microscopes themselves.

The discussion of image contrast in electron microscopy and X-ray topography (chapters 5 and 6 and part of chapter 2) is inevitably in terms of diffraction theory. No knowledge of dynamical theory of diffraction is assumed, but we have not developed diffraction theory from the very beginning since it will certainly be familiar to the majority of readers. The main knowledge required is of the geometric (kinematical) ideas of diffraction of waves by crystals, in particular the Bragg law, the reciprocal lattice representation and the Ewald sphere construction. These will be used extensively, but all the necessary information can be found in one or two chapters of books such as Ball (1971) or Barrett and Massalski (1966). These books also cover the elementary crystallography that is needed for chapters 5, 6 and 7 and part of chapter 2—essentially an acquaintance with the simpler crystal structures and the use of Miller indices for crystal planes and directions. Crystallographic group theory is not used.

We have taken considerable trouble to formulate the mathematics simply. In general, only school-level algebra and calculus are used, this mainly in chapter 5 (the results of which are used in chapter 6). Chapter 5 also contains reference to elementary Fourier theory, and the noise calculation in chapter 2 (also used in chapters 3 and 6) uses a little statistics; these topics are not extensively employed and could be taken on trust, but an excellent introduction to Fourier theory can be found in Lipson and Lipson (1969), while texts such as Pugh and Winslow (1966) cover the statistics required (and much more). Vector notation and simple vector algebra are frequently used, and can be quickly acquired from a standard book on mathematics or from a text such as Feynman *et al.* (1963).

Since the scope of this book is very wide there are, inevitably, references to some aspects of modern physics with which the more technically trained reader may be unfamiliar, such as the discussion of the scattering processes in solids and some of the applications of the techniques, like the study of electronic devices or of magnetism. Again, these may be taken on trust or ignored by those interested only in some other specific application (for example, the mechanical engineer will not require quantum theory to understand and use the scanning electron microscope for looking at wear on bearings), but readers interested in following up any of these topics will find the books by Feynman *et al.* (1963) very valuable.

Finally, it may be useful to indicate some reviews of the topics covered in section 1.1. There are many good books available on the optical microscopy of materials but a valuable source for this and many other techniques is Chalmers and Quarrell (1960), while Tolansky (1973) is excellent on interferometry. X-ray diffraction methods are treated in some detail by Cullity (1956), by Barrett and Massalski (1966) (who also discuss electron and neutron diffraction) and by Guinier (1952).

REFERENCES

Ball, C. J., (1971), *An Introduction to the Theory of Diffraction*, Pergamon, Oxford

Barrett, C. S., and Massalski, T. B., (1966), *Structure of Metals*, 3rd edn, McGraw-Hill, New York

Chalmers, B., and Quarrell, A. G., (1960), *The Physical Examination of Metals*, (eds.) 2nd edn, Arnold, London

Cullity, B. D., (1956), *Elements of X-ray Diffraction*, Addison-Wesley, Reading, Mass.

Feynman, R. P., Leighton, R. B., and Sands, M., (1963), *The Feynman Lectures on Physics*, Addison-Wesley, Reading, Mass.

Guinier, A., (1952), *X-Ray Crystallographic Technology*, Hilger and Watts, London

Lipson, S. G., and Lipson, H., (1969), *Optical Physics*, Cambridge University Press

Pugh, E. M. and Winslow, G. H., (1966), *The Analysis of Physical Measurements*, Addison-Wesley, Reading, Mass.

Reed-Hill, R. E., (1973), *Physical Metallurgy Principles*, 2nd edn, Van Nostrand, New York

Tolansky, S., (1973), *An Introduction to Interferometry*, Longman, London; also (1970), *Multiple Beam Interference Microscopy of Metals*, Academic Press, London

Van Vlack, L. H., (1970), *Materials Science for Engineers*, Addison-Wesley, Reading, Mass.

2

Scanning Electron Microscopy

The scanning electron microscope (SEM) is an instrument designed primarily for studying the surfaces of solids at high magnification. In this respect it may be compared with the optical microscope, and a set of micrographs taken on each instrument is shown in figure 2.1. The specimen is an aluminium alloy broken by ductile fracture, and it is obvious that the images are qualitatively similar but that the SEM possesses much greater resolution and depth of field. If this were all, then the SEM would still be a most useful microscope and indeed a very large number of its applications utilise only these advantages. But there are additional benefits which arise from using electron beams rather than light beams in image formation since the interaction of electrons with solids is more diverse than that of photons, and correct use of the SEM can yield much more information—for example, on crystal orientation, chemical composition, magnetic structure or electric potential in the specimen.

2.1 PRINCIPLES OF OPERATION

Images are formed in the SEM by a quite different mechanism from that in an optical microscope. No objective lens is used, but instead, images are built up point by point, in a way similar to that used in a television display. A fine, high-energy beam of electrons is focused to a point on the specimen. This causes the emission of electrons (with a wide spread of energies) from that point on the surface, and the emitted electrons are collected and amplified to give an electrical signal. If this signal is now used to modulate the intensity of a beam of electrons in a cathode-ray tube (CRT) display, one point of the image is formed on the CRT. To build up the complete image, the electron beam in the microscope is scanned over an area on the specimen surface (the pattern of scan is called a raster) while the beam in the CRT display is scanned over a geometrically similar raster. Thus when the scanning beam reaches the part of the specimen corresponding to A in figure 2.1a, very few electrons are emitted and the intensity of the synchronously scanned beam in the CRT is low; at B, the emitted intensity increases greatly and the CRT beam is

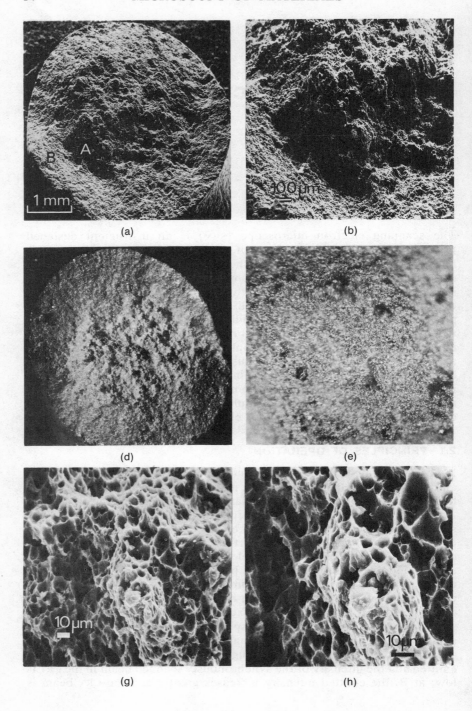

(a)

(b)

(d)

(e)

(g)

(h)

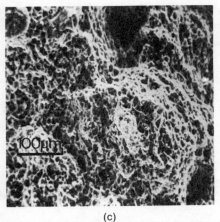

(c)

Figure 2.1. Fracture surface in aluminium alloy, broken by ductile fracture. (a), (b) and (c) are scanning electron micrographs, (d), (e) and (f) are optical micrographs at corresponding magnifications; (g), (h) and (i) are further electron micrographs at higher magnification

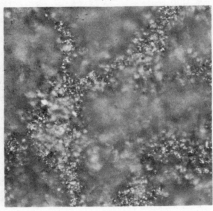

(f)

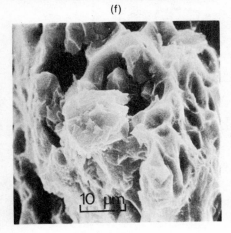

(i)

made proportionately more intense. The image on the CRT is thus a map of the intensities of the electron emission from the specimen surface, just as the image in a metallurgical microscope is a map of the light reflected from the surface. The system is closely analogous to a closed-circuit television system, in which the 'specimen' is the image formed by an optical lens on a special screen. This screen is scanned by an electron beam and the resulting signal used to modulate the beam in a television monitor. Again, the rasters described by the electron beams in the camera and monitor must be geometrically similar and in exact synchronism. Figure 2.2 shows block diagrams of the two systems and it is seen that geometrical similarity and synchronism of the two electron beams is in each case achieved by using the same electronic unit (the 'scan

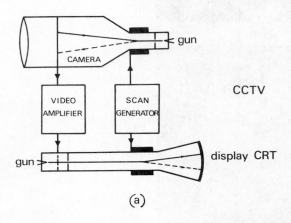

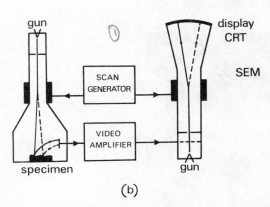

Figure 2.2. Block diagrams of systems employing electron beam scanning: (a) closed-circuit television, (b) scanning electron microscope. (Courtesy of C. G. van Essen)

generator') to control each beam. In an SEM the raster is normally square and is covered in a line-by-line scan of up to 1000 lines (as compared with 625 lines in most television sets).

The magnification is given simply by the ratio of the side-lengths of the display and specimen rasters, and is normally variable from about 20× to 100000×. The resolution, however, is the more important quantity for any microscope, and with an ideal specimen it is at best equal to the diameter of the electron beam where it strikes the specimen surface (the beam cannot in practice be focused to a perfect point). In current high performance instruments, this can be as small as 5 nm, compared with a resolution of about 300 nm for an optical microscope. However, the resolution in the SEM depends critically upon the nature of the specimen and the mode of operation of the instrument, and 15 nm is a more typical figure.

2.2 SPECIMEN PREPARATION

One of the many merits of the SEM is the ease with which specimens can be prepared, especially for electrically conducting materials. These are merely cleaned and stuck onto a small aluminium mounting stub with electrically conducting glue (figure 2.3). Specimens up to a few mm in each dimension can easily be accommodated, and larger specimens can be used with some loss of movement. If the object cannot be inserted in the microscope (for example, the cylinder of a combustion engine), then a plastic replica can be made and treated as for insulating specimens.

The difficulty with electrically insulating materials is that they accumulate electrical charge from the primary electron beam. This charge deflects both the beam and the trajectories of the collected electrons and a grossly distorted image results. The specimen can either be examined at a

Figure 2.3. Scanning electron microscope specimens on their mounting stubs (which are 10 mm diameter)

reduced accelerating voltage, when the electron emission from the surface can balance the rate of arrival of electrons in the beam, or it may be coated with a thin layer of conducting material.

Two coating methods are commonly used. In the first, a layer of one of the 'anti-static' preparations is applied; these are long-chain organic molecules developed empirically in the textile industry, which suppress the accumulation of charge. Their action is poorly understood (a fact that has limited their more widespread use) but they may in some cases be effective when only a single molecular layer is present. In the second method, a layer of carbon, silver, gold or similar conductor about 20 nm thick, is applied in a conventional vacuum evaporator. It is difficult to coat complex shapes perfectly uniformly and the full resolution of the SEM cannot, in general, be employed, but the majority of work is at a magnification around 5000× for which coating procedures are quite satisfactory. The problems are most acute for biological specimens, which are also more susceptible to damage from radiation, heating and the high-vacuum environment.

Texts such as Hearle *et al.* (1972), Oatley (1972) and the *I.I.T. Symposia* (see section 2.8) should be consulted for full details of specimen-preparation techniques.

2.3 DESIGN AND CONSTRUCTION OF THE MICROSCOPE

Figure 2.4 shows a schematic diagram of a typical SEM (details of the arrangement vary between different manufacturers). The essential features of the microscope are

(1) An electron gun, to produce a narrow beam of electrons accelerated through a potential difference of up to about 50 kV

(2) Two or three lenses to focus the electron beam as finely as possible

(3) A system to deflect the beam over the raster on the specimen

(4) A specimen stage permitting movement, tilt and rotation of the specimen

(5) A method of collecting and amplifying the emitted electrons

(6) Cathode-ray tubes to display the image

(7) Electronic circuits to supply and control the electron gun, accelerating voltage, lens currents, scan generator and signal amplification

(8) A high-capacity vacuum system to maintain a pressure below 10^{-5} torr and permit rapid evacuation after changing the specimen

Since some of these requirements are common to several types of electron-optical instruments, the following sections will be used throughout the book.

2.3.1 The Electron Gun

The source of electrons is usually a hairpin-shaped tungsten filament (as used in a television set) which is heated to emit electrons by thermionic

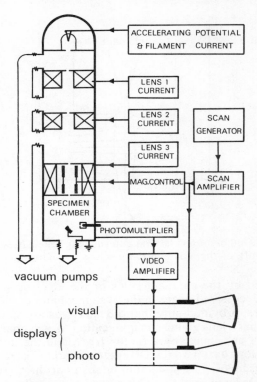

ACCELERATING POTENTIAL
& FILAMENT CURRENT

LENS 1
CURRENT

LENS 2
CURRENT

SCAN
GENERATOR

LENS 3
CURRENT

MAG.CONTROL

SCAN
AMPLIFIER

SPECIMEN
CHAMBER

PHOTOMULTIPLIER

vacuum pumps

VIDEO
AMPLIFIER

visual

displays {

photo

Figure 2.4. Schematic diagram of a scanning electron microscope. The details (for example, the number of lenses) vary between different manufacturers

emission. Figure 2.5 shows a suitable electron gun. The metal cap ('Wehnelt cylinder') over the filament, together with the anode, form a simple electrostatic lens. The Wehnelt cylinder is usually biased to act as a grid. The curvature of the electric field focuses the electrons to the 'first crossover point'; at this point, an image of the filament tip (cathode) is formed, about 60 μm in diameter. From the anode onwards the electrons are in free flight down the microscope column; using an accelerating voltage of 30 kV, they will be travelling at about one-third of the velocity of light.

It is useful to distinguish between the intensity and the brightness of the beam produced by an electron gun plus a system of lenses. The intensity I is the current density, measured in amps per square metre; the brightness B is the current density per unit solid angle, measured in amps per square metre per steradian, and the two are related at an image plane by the expression

$$I = B\alpha^2\pi \tag{2.1}$$

where α is the 'beam divergence' (the semi-angle of the cone of rays converging to the image). It is the intensity that determines the exposure time required to record an image; but the brightness is a useful parameter

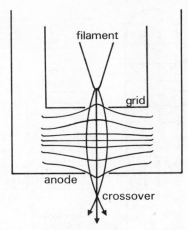

Figure 2.5. Outline of an electron gun showing schematically the equipotentials in the vicinity of the axis and the electron trajectories

since it is fixed by the characteristics of the electron gun and remains constant for successive images in the system whatever the magnifications or the apertures, provided only that the apertures are small. This follows from the fact that a lens giving a magnification M will change the intensity between object and image by a factor $1/M^2$, and the divergence by the same factor as long as $\tan \alpha \approx \alpha$. Divergences rarely exceed 10^{-2} radians in electron microscopes. The maximum intensity at the final image is given by multiplying the gun brightness by $\pi \alpha^2$ where α is the divergence at the image (controlled by the final aperture). This intensity is of course less than that at the first image of the filament just after the electron gun since not all the initial divergence is used, as shown in figure 2.10a. The ratio of the number of electrons that form one image to the number that are accepted by the subsequent lens is approximately $(\alpha_2/\alpha_1)^2$, where α_1 and α_2 are the divergences accepted by the lenses before and after the image, respectively.

The usual tungsten hairpin-filament gun has a brightness limited by its temperature and by the nature of thermionic emission. Since brightness is so important for reducing noise in the image (section 2.4.2) and for making rapid dynamic studies, several less conventional electron sources have been investigated. These include pointed tungsten filaments or oxide or boride crystals (in particular lanthanum hexaboride), run under thermionic or field-emission conditions (field emission occurs when electrons are pulled out of the solid by the intense electric field acting at a sharply pointed tip; a true field-emission source will operate at room temperature and is very bright). The thermionic pointed filaments are not so good as field-emission sources but are an improvement on hairpin filaments since the emitting area is smaller. Some typical brightness values are given in table 2.1 and it is clear that the newer guns are substantially better, though they do require higher vacua. Lanthanum hexaboride guns

Table 2.1. Typical brightness values and source sizes for various types of electron gun

Type of gun	Brightness ($A\ mm^{-2}\ sr^{-1}$)	Source diameter (μm)
Tungsten hairpin	10^2-10^3	50
Tungsten point, thermionic ⎱ Lanthanum hexaboride ⎰	10^3-10^4	1–10
Tungsten point, field emission	10^6	0.01–0.1

are now a standard option on several models of SEM, pointed filaments can be obtained for almost any model; and microscopes with field-emission guns are available despite the extra expense and complexity caused because field-emission sources only work in an ultra-high-vacuum environment. High-brightness guns are also useful in the transmission electron microscope for high-resolution work, principally because the more intense image requires a shorter exposure time, thus minimising the problems of electrical and mechanical instability.

2.3.2 The Electron Lenses

The purpose of the lens system is to form as small an electron probe as possible, that is, to produce a greatly demagnified image of the cathode. It is possible to use electrostatic lenses, which (as in the electron gun) consist essentially of sets of plates containing holes to permit the passage of the beam, the plates being held at different potentials; the curvature of the electric field produces a focusing action. Such lenses, however, can be rather troublesome since they need careful electrical insulation against the high potentials required. They also suffer from large aberrations, and modern microscopes use magnetic lenses at all stages after the electron gun.

The force on a charged particle moving in a magnetic field is given by the vector equation $F = qv \times B$, where F is the force, q the charge, v the velocity of the charge and B the magnetic induction. The direction of the force is therefore perpendicular to both the magnetic induction and to the velocity of the charge, and lenses must use fields that have components perpendicular to the velocity of the charge. In addition, any field used for a lens must be axially symmetric. A simple current-carrying coil satisfies these conditions, and this type of lens is used in cathode-ray tubes, for which aberration problems are not serious. An electron microscope lens is essentially a pair of cylindrical magnet poles with circular coaxial bores (figure 2.6); the magnetic circuit is completed and the whole lens energised with a coil (carrying a current typically in the range 50 mA–1 A) as shown in figure 2.7.

The vector B can be represented by two components, B_z along the axis of the lens and B_r in the radial direction (figure 2.8). The paths of

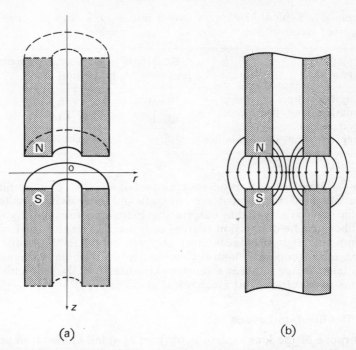

(a)   (b)

Figure 2.6. The basis of a magnetic lens: (a) a pair of cylindrical magnet poles with circular coaxial bores, (b) axial cross-section showing magnetic field in the vicinity of the gap (after Hall, 1966)

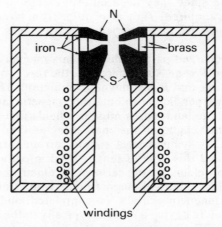

Figure 2.7. Cross-section of typical strong magnetic lens. The polepiece (shown in black) serves to concentrate the field and hence make the focal length short; a lens might have several interchangeable polepieces

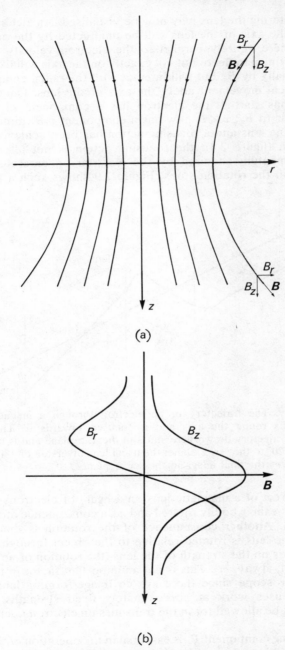

(a)

(b)

Figure 2.8. (a) Magnetic field (induction) lines near the axis of a lens (compare figure 2.6b), (b) the axial and radial components of the magnetic induction, B_z and B_r, near the axis (after Hall, 1966)

electrons entering the lens may now be visualised. An electron travelling exactly on the axis of the lens will be unaffected by the magnetic field since it is here entirely parallel to the electron velocity ($B_r = 0$). An electron moving parallel to but not exactly on the axis will likewise not be affected initially by B_z, but will interact with the radial component B_r to cause a helical movement about the axis of the lens. Once this helical movement has started, the electron has a component of velocity, v_t, perpendicular to B_z; $v \times B$ now has a component $v_t B_z$ directed towards the axis of the lens and a focusing action has been achieved. Finally B_r changes sign (figure 2.8); the focusing action is not affected but the angular momentum acquired by the electron on entering the lens is cancelled and the rotation stops. Figure 2.9 shows such a trajectory.

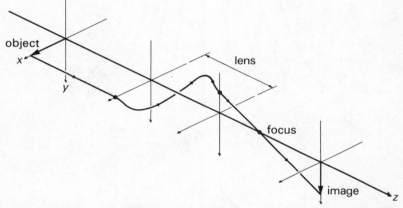

Figure 2.9. The trajectory of an electron through a magnetic lens; the electron spirals round the axis and is focused towards it. The (somewhat exaggerated) trajectory drawn here starts in the x, z plane and is rotated by the lens through 270° to the $-y$, z plane; the usual lens inversion of 180° through the focal point gives the final image in the $+y$, z plane

The net effect of a magnetic lens on a beam of electrons is therefore twofold: the beam is bodily rotated and, as a consequence of the rotation, it is focused. Another consequence of the rotation is that any image formed by the lens is rotated relative to the object through some angle which depends on the strength of the lens (the rotation of an optical lens is, in contrast, always π). This is not an important factor in the scanning electron microscope since there are no image-forming lenses and the condenser lenses work at approximately fixed strengths, but image rotation must be allowed for in the transmission electron microscope (see chapter 4).

Although the component B_r is essential in the operation of the magnetic lens, it is possible to derive equations for the focal length, f, in terms of B_z alone (see Hall, 1966). An approximate expression for a thin magnetic lens is (in SI units)

$$\frac{1}{f} = \frac{e}{8mV} \int_{\text{lens}} B_z^2 \, dz \qquad (2.2)$$

where m is the electronic mass and V the accelerating voltage prior to the lens, and although actual lenses may be too 'thick' to obey this formula, it does illustrate some general points which hold for all magnetic lenses.

(1)   The focal length depends on the e/m ratio of the particles (this is in contrast to the behaviour of electrostatic lenses).

(2)   The accelerating voltage affects the focal length, through control of the electron velocity* (faster electrons are more difficult to deflect). Thus high stability of the accelerating voltage (to about 1 part in 10^5) is essential, to reduce this 'chromatic aberration'.

(3)   Whatever the sign of the magnetic induction, the focal length is positive. This has important consequences, since it is the usual practice in optics to correct lens aberrations using divergent lenses (f negative) of different refractive index. Divergent magnetic lenses are not possible, a fact which seriously limits the resolution of electron microscopes. A similar conclusion holds for electrostatic lenses if the potential is the same each side of the lens. It is possible to make divergent electron 'lenses' that are not axially symmetric but these obviously have limited uses in microscopy.

(4)   The focal length of the lens is shorter the higher the magnetic induction B_z. Variation in the excitation of an electromagnetic lens hence results in a change of focal length. This is very convenient for focusing the images and (in the transmission microscope) for altering the magnification, but means that for high resolution the lens currents must be highly stable, usually to 1 part in 10^6.

Magnetic lenses are by no means perfect, and chromatic aberration has already been mentioned. Their most serious defect, however, is spherical aberration; rays are focused more strongly the greater their inclination to the axis. This aberration reduces as the magnetic induction is increased, and strong lenses are used for high performance. The focal lengths of such lenses, as used in the final stage of the SEM and as the objective lens of the transmission microscope, range from about 1 to 25 mm. Electron lenses are also (like human eye-lenses) prone to astigmatism, which causes a point object to be imaged as a line. This arises from imperfections in manufacture or from dirt in the microscope column, and correctors are provided to reduce or eliminate its effect upon the resolution. It may be noted here that resolution in the scanning electron microscope is usually limited by the nature of the specimens rather than by the perfection of the lenses. This is not so in the case of transmission electron microscopes, and lens aberrations will be discussed more fully in chapter 4.

*Electrons in electron microscopes are travelling at such high velocities that relativistic effects are significant. It is not worth including them in this crude equation for the focal length, but for calculation of the de Broglie wavelength, for example, relativistic effects contribute a correction of about 5 per cent after acceleration through 100 kV.

2.3.3 The Scanning System

Ideally, one would place the scanning system after the final lens in the SEM, but there is insufficient space in this position because the final lens must have a short focal length in order to reduce its aberrations. The beam is therefore deflected through the required angle before the final lens, and given a further deflection so that it will pass through the centre of the lens. Figure 2.10a shows the ray path for the complete microscope, including this double deflection system.

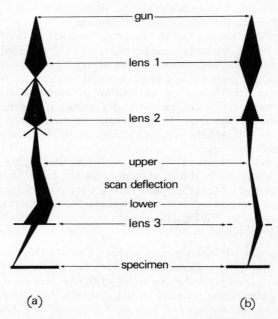

gun

lens 1

lens 2

upper

scan deflection

lower

lens 3

specimen

(a)   (b)

Figure 2.10. Ray paths in the SEM: (a) standard arrangement for image formation, (b) arrangement for selected-area channelling patterns, section 2.5.7. (Courtesy of C. G. van Essen)

The deflection system consists of two sets of coils which produce a magnetic field perpendicular to the microscope axis as shown in figure 2.11. The deflection is of course perpendicular to both the field and the axis. The coils should have reasonably low impedance so that the scan can be made very rapid if required.

It will also be seen from figure 2.10a that apertures are included at various points to limit the divergence of the beam (these are often called 'spray' or 'splash' apertures). Extra sets of deflection coils, arranged to displace the beam laterally without altering its inclination to the axis, are sometimes provided to align the beam relative to these apertures and the lenses.

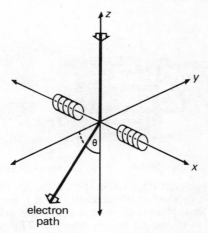

Figure 2.11. Principle of operation of deflection coils for scanning or beam alignment. The coils produce a field along the $+x$ axis; the electron, travelling in the $-z$ direction, is deflected in the $-y$ direction while remaining in the yz plane. Several designs of coils in combination with iron cores are possible (see Oatley, 1972)

2.3.4 The Specimen Stage

The requirements for the specimen stage are that it should be mechanically stable, accept as large a specimen as possible, provide movements in three perpendicular directions for looking at different areas of the specimen and for adjusting the working distance between the specimen and final lens, provide rotation and tilt of the specimen (since these adjustments affect the contrast of the image), and provide electrical and other feedthroughs for performing experiments on the specimen within the microscope. Specimen chambers are usually quite large and the satisfaction of these requirements is a straightforward problem in mechanical design. An example is shown in figure 2.12. It should be emphasised that the use of the microscope as an experimental chamber is a major factor in the importance of the SEM, and special stages have been constructed for observations at high temperatures, very low temperatures, during mechanical testing, under chemical attack by corrosive gases, and under the influence of magnetic or electric fields (to name but the most common; for a review see Hearle *et al.* 1972).

2.3.5 The Electron Collector

The design of an efficient electron collector by Everhart and Thornley (1960) is usually regarded as the turning point in the design of a successful SEM. Their design is shown in figure 2.13. The grid at the front can be biased by ± 250 V relative to the specimen, which is at earth potential. As discussed later, the electron emission from a specimen falls into two main

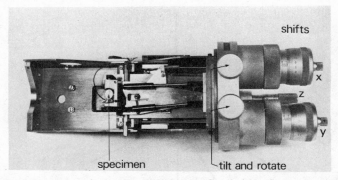

Figure 2.12. A specimen stage (from the Cambridge Stereoscan). The functions of the controls are shown: the electron beam is assumed to be in the *z* direction and the tilt-and-rotate controls can incline the specimen surface at any angle and orientation with respect to this direction. The electron collector (not shown) is positioned at the end of the stage opposite from the controls

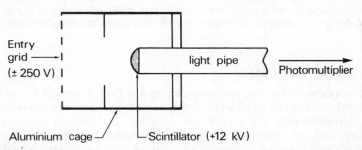

Figure 2.13. The Everhart–Thornley design of electron collector. The grid voltage is held at about +250 V to attract secondary electrons and −250 V to repel them. The potential of +12 kV on the scintillator is to ensure that electrons within the cage are attracted to the scintillator and strike it with sufficient energy

regions, high-energy ('primary') electrons with approximately the energy of the original electron beam (e.g. 30 keV) and 'secondary' electrons with energies below about 50 eV. The bias potential on the grid can therefore be used either to pull as many secondary electrons into the collector as possible (positive bias) or to repel them completely (negative bias) so as to allow only the high-energy electrons to participate in image formation. The selected electrons pass through the grid and enter a Faraday cage; they are then strongly attracted towards a plastic scintillator coated with a thin layer of aluminium which is maintained at a potential of about +12 kV. Each electron produces a flash of light on striking the scintillator; the light is conducted outside the vacuum chamber by means of a perspex light pipe and enters a photomultiplier, where it is converted back into an electrical signal and amplified. This system produces a noise-free amplification of up to $10^6\times$ which is virtually 100 per cent efficient and has a bandwidth of 10 MHz. The signal may now be amplified conventionally and used to modulate the CRT display. It is worth noting that this apparently indirect method of amplification is much superior in gain, speed of response and freedom from noise to more conventional methods, mainly because of the excellent characteristics of photomultipliers. The current flowing to earth through the specimen is complementary to the emitted current (though not all emitted electrons are collected as will be seen later) and so can be used to modulate the CRT to form the image. This 'absorptive' mode of operation has some advantages but has in the past been used less than it might have been because the associated 'specimen-current amplifiers' have been slower and more temperamental than the Everhart–Thornley detector. However, improvements in electronics now appear to have solved the problem and currents as low as 10^{-13} A or less can be detected quite satisfactorily.

It is usual to record the image by a single slow scan, taking perhaps 40 or 100 s, in order to minimise the noise on the image (section 2.4.2). A conventional CRT with a short-persistence phosphor is used to display the image for photography by a roll-film or Polaroid camera. The short-persistence phosphor is required to avoid blurring of bright features on the image, but for visual observation a longer-persistence phosphor will reduce noise by integrating the image between frames, which can then be as frequent as once a second. If a high-brightness gun is available or only a low resolution is needed, then television scan rates can be used (when again a short-persistence phosphor must be used on the CRT). In either case, it is usual to provide separate CRTs for visual observation and photography, and some instruments have separate channels, each with its own amplifiers and CRTs, so that several contrast modes may be used at once. It is also possible to provide high and low magnifications simultaneously on separate screens.

2.3.6 The Scope of Commercial Microscopes

The most advanced research microscopes are capable of operating in almost all of the modes described in this chapter. They are invaluable items of equipment for a research laboratory, especially one that habitually tackles a wide range of problems. However, they are scarcely suitable for a teaching laboratory (if only because of their cost), or for routine process control when only one type of image contrast is required. Indeed, the point has already been made that a high proportion of the applications of the scanning electron microscope arise solely from its high resolution and high depth of field. This has led to the production of simple instruments at much lower cost than the research microscopes. In these 'cheap' microscopes (which might cost about the same as a very elaborate optical microscope) one would have an acceptable resolution, perhaps even as good as in the advanced models; high depth of field of course; probably a large specimen chamber so that production components could be examined; a design optimised for convenience of use (for example, TV scanning rates would be employed if possible); and a high throughput of specimens. On the other hand, the more specialised contrast modes described in section 2.5 would not be available (such as voltage, cathodoluminescent or crystallographic contrast) and there would be little if any possibility of conducting experiments on the specimen inside the microscope.

2.4 THE PERFORMANCE OF THE SEM

Before embarking on the details of image contrast, and the applications of the microscope, it is useful to discuss its capabilities and limitations: how well can the instrument perform given an ideal specimen, and how can the nature of the specimen limit the performance? The important characteristics are those of depth of field (a purely instrumental factor), image noise (on which the specimen has some influence), and resolution (on which the specimen has a very great influence).

2.4.1 Depth of Field

This is the distance along the microscope axis through which the specimen can be moved without perceptibly blurring the image (it is sometimes wrongly called depth of focus). Figure 2.14 illustrates this

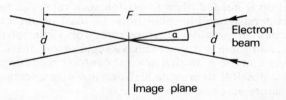

Figure 2.14. Calculation of the depth of field. The electron beam is shown converging to the image plane with semi-angular aperture α, d is the resolution required and F the depth of field at this resolution

situation for a fixed beam divergence α. Blurring is measured by the diameter of the 'disc of confusion', and it is seen that a disc of confusion of diameter d is related to the axial shift F by the equation

$$\tfrac{1}{2}F \tan \alpha = \tfrac{1}{2}d \tag{2.3}$$

or

$$F = d/\alpha \text{ for small } \alpha \tag{2.4}$$

F is equal to the depth of field when the disc of confusion is no larger than the resolution obtained at the particular magnification, M, which is in use. For an image 100 mm square formed with a raster of 1000 lines, the resolution is $0.1/M$ in millimetres. Then $d = 0.1/M$ and

$$F = \frac{0.1}{M\alpha} \tag{2.5}$$

The depth of field at various magnifications (and hence resolutions) is given in table 2.2, for a typical aperture of 5×10^{-3} radians. The table also

Table 2.2. Depth of field and resolution of the SEM *(final aperture 5×10^{-3} rad) and the optical microscope*
(The highest resolutions cannot always be attained, as discussed in section 2.4.3)

Magnification	Resolution	Depth of field SEM	Optical
20	5 μm	1 mm	5 μm
100	1 μm	200 μm	2 μm
200	500 nm	100 μm	0.7 μm
1000	100 nm	20 μm	—
5000	20 nm	4 μm	—
10000	10 nm	2 μm	—

gives figures for an optical microscope. Equation 2.3 applies equally to this instrument but the apertures are much larger and the depth of field consequently one or two hundred times less than in the SEM. Because a high depth of field is so useful, the final aperture is often chosen simply to give a sufficient depth of field, but for high resolution an optimum aperture must be used, as explained in section 2.4.3.

2.4.2 Noise

The noise problem is that of collecting enough information to form an image. It arises in all forms of microscopy because the processes used to generate the illuminating radiation, such as thermionic emission of electrons, thermal emission of photons or characteristic X-ray emission, are themselves random processes and, in general, so are the scattering processes in the specimens. This affects the resolution, since two adjacent points will only be resolved if the difference between the signals they emit is greater than the statistical uncertainty in those signals. The resolution will be improved if the signal from the two points is collected for a longer time, and also if the contrast between the points is increased; a larger signal means a lower fractional error in the signal. Alternatively, areas of feeble contrast can be seen if they are large (that is, if resolution is sacrificed) since the signal is obviously proportional to the area.

These points are illustrated in figure 2.15 where a was taken under proper operating conditions but b was taken with the primary beam current considerably reduced. The so-called 'shot-noise' in the image is very evident. The inter-relation between contrast and resolution can also be seen: features with high contrast or of large size are visible in both pictures but the smaller, fainter details visible in a are difficult to see or even obscured by the noise in b (see for example the circled regions in each photograph).

Although the terminology of the scanning electron microscope has been used above, it is worth repeating that this noise problem is fundamental,

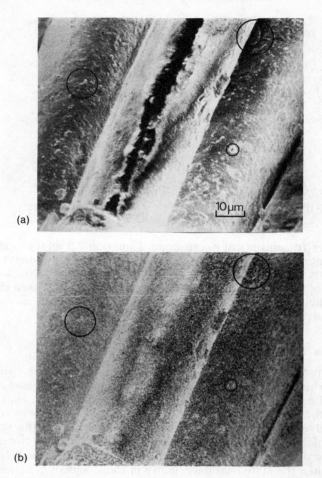

Figure 2.15. The loss of image detail through shot-noise (statistical noise). The specimen is a bundle of nylon fibres from a nylon glove; (a) correct operating conditions, (b) beam current reduced to increase the noise. The circles surround corresponding features on each micrograph

not depending on the make of instrument or even on the type of microscope. For example, the ideas and calculations of this section will be used again in section 6.1.4, in the context of X-ray topography. It is therefore useful to be able to calculate the effects of statistical noise on the resolution, given the illumination and the contrast levels.

Consider two adjacent areas each of diameter d, where d is the resolution that is desired. Let the signal from one area be S_1 and its standard error be σ_1, and let S_2 and σ_2 be the corresponding parameters from the other area. We shall assume a normal distribution of signal strengths from both areas. Now, we need to be confident that S_1 is not S_2.

It is a well-known property of the normal distribution that there is a 99.7 per cent probability that the true mean of a set of readings lies in the range $M \pm 3\sigma_M$ where M is the mean of the set and σ_M the standard error (standard deviation of the mean). To this degree of confidence then, the required condition is

$$S_1 + 3\sigma_1 \leqslant S_2 - 3\sigma_2 \tag{2.6}$$

assuming that $S_1 < S_2$, as illustrated in figure 2.16. The contrast, C, is

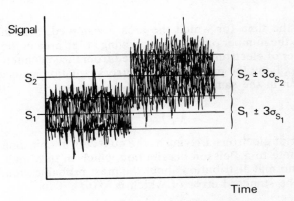

Figure 2.16. The criterion used to distinguish between two noisy signals that differ by a contrast C [that is, $S_2 = S_1(1 + C)$]; the 99.7 per cent confidence limits of the two signals just touch

defined as the change of signal divided by the original signal, so $S_2 = S_1(1 + C)$ and equation 2.6 becomes

$$S_1 + 3\sigma_1 \leqslant S_1(1 + C) - 3\sigma_2 \tag{2.7}$$

or

$$S_1 C \geqslant 3(\sigma_1 + \sigma_2) \tag{2.8}$$

Clearly, when noise is present a higher signal is required for regions of weak contrast. It is convenient to approximate $\sigma_1 \approx \sigma_2 = \sigma_S$; the criterion then becomes

$$C \geqslant 6\left(\frac{\sigma_S}{S}\right) \tag{2.9}$$

This approximation is on the safe side, since $S_1 < S_2$ and thus $\sigma_1 > \sigma_2$; however, the effect is quite small (even for a contrast of 100 per cent the coefficient 6 in the last equation would only reduce to about 4.5). The problem now becomes that of deciding the value of the standard error σ_S. The signal reaching the final image is a product of i the electron current leaving the specimen, p the efficiency of the collecting system, and g the gain of the amplification system, thus

$$S = ipg \tag{2.10}$$

The appropriate statistical equation for the propagation of errors gives the fractional error σ_S/S as

$$\left(\frac{\sigma_S}{S}\right)^2 = \left(\frac{\sigma_i}{i}\right)^2 + \left(\frac{\sigma_p}{p}\right)^2 + \left(\frac{\sigma_g}{g}\right)^2 \tag{2.11}$$

σ_p may be taken as zero since the efficiency of collection depends on geometrical factors and the (presumably fairly stable) collector voltage. The current leaving the specimen is given by

$$i = \frac{(nf)e}{t} \tag{2.12}$$

where t is the time for which the area is sampled, e is the electronic charge, n is the number of electrons arriving at the area in the time t and f is the number of electrons that are emitted from the specimen per incident electron. Applying the propagation-of-errors equation again and ignoring any variation in the sampling time

$$\left(\frac{\sigma_i}{i}\right)^2 = \left(\frac{\sigma_{nf}}{nf}\right)^2 \tag{2.13}$$

The number of electrons arriving at the collector in unit time is expected to approximate to a Poisson distribution which in turn may be approximated to a normal distribution about the most probable value (as long as $(nf) \geqslant 10$), the standard error of which is $\sqrt{(nf)}$. Thus

$$\left(\frac{\sigma_S}{S}\right)^2 = \frac{1}{nf} + \left(\frac{\sigma_g}{g}\right)^2 \tag{2.14}$$

and, from equation 2.9

$$C^2 \geqslant 36\left\{\frac{1}{nf} + \left(\frac{\sigma_g}{g}\right)^2\right\} \tag{2.15}$$

The resolution can now be related to the contrast by evaluating n. This is given by multiplying the spot intensity, equal to $B\alpha^2\pi$ from equation 2.1, by the spot area $\pi d^2/4$ and the time t and dividing by e. Replacing t by τ/N^2 where τ is the time to scan a whole frame and N is the number of lines per frame

$$n = \frac{Bd^2\alpha^2\tau\pi^2}{4N^2e} \tag{2.16}$$

Substituting this expression in equation 2.15 and rearranging to show the effects of the parameters on the resolution, d

$$d^2 \geqslant \frac{144N^2e}{\alpha^2\tau Bf\pi^2[C^2 - 36(\sigma_g/g)^2]} \tag{2.17}$$

This clearly shows the effects. N is normally fixed at 1000 lines, B is fixed by the gun used and f by the specimen and the mode of contrast chosen. The resolution improves with an increased time τ for one frame scan, with increased contrast in the specimen and with an increased value of the final aperture size (though if the latter is increased too far the resolution will be limited by aberrations in the objective lens). Noise in the amplification system will also degrade the resolution.

The noise-limited resolution has been derived in a fairly general way to bring out the important features, and to allow the equations to be used for other forms of microscopy. The only approximation that cannot always be made is that the rate of emission of electrons from the specimen follows a Poisson distribution (which gives a simple expression for the standard error). Thermionic emission from the filament should certainly follow Poisson statistics, but the scattering processes in the specimen may destroy this distribution and, depending on the precise nature of the beam/specimen interactions, σ_{nf} may in practice be greater or less than $\sqrt{(nf)}$. If the interactions are well known, the equation for the propagation of errors may again be applied to give

$$\left(\frac{\sigma_{nf}}{nf}\right)^2 = \left(\frac{\sigma_n}{n}\right)^2 + \left(\frac{\sigma_f}{f}\right)^2 \tag{2.18}$$

and the other equations may be modified accordingly. However, these interactions are not often known in such detail, and sufficient accuracy is obtained in practice by using the semi-empirical formula obtained by Oatley *et al.* (1965) who in effect increased the factor $144/\pi^2$ to 100 to take account of both these variations and also the noise in the amplification system (strictly, this is not linear with the other parameters as seen from equation 2.15 but is so small that the result certainly errs on the cautious side). Their equation (with the factor f inserted to cope with situations such as X-ray emission which are very inefficient) is

$$d^2 \geqslant \frac{100N^2e}{\alpha^2\tau BfC^2} \tag{2.19}$$

This equation is shown in figure 2.17 as a plot of the minimum resolution against the specimen contrast for different frame times, using parameters appropriate to a conventional hairpin-filament gun. Long scan times are needed for high resolution with specimens of low contrast, and it is clearly helpful to use the maximum beam diameter consistent with the resolution needed, to minimise the noise. At $100\times$ magnification for example, a probe diameter of 15 nm is quite unnecessary and a 1 μm diameter spot can be used to advantage. The difficulties of using television-rate displays are also brought out by the graph; at the corresponding frame scanning time of 40 ms the resolution is poor even with a contrasty feature, and the value of the high-brightness guns discussed in section 2.3.1 is evident. The slow scanning rates normally used (especially for photographing the image) are seen to be a direct consequence of the inherent noise in the imaging system and in the specimen itself.

2.4.3 Resolution

If a specimen with sufficient contrast is available so that small probe diameters can be used with reasonable frame scan speeds then the resolution will be governed by the performance of the instrument (provided that emitted electrons come only from the area struck by the probe). This instrumental resolution will be governed by the balance

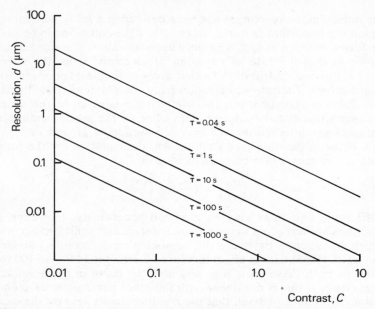

Figure 2.17. The minimum electron beam diameter (best resolution) plotted against the contrast for different frame times τ, calculated from equation 2.19, to show the effects of noise on resolution. Parameters typical of a conventional tungsten hairpin-filament gun have been used

between the effects of the aberrations of the final lens (which increase with lens aperture) and of diffraction effects (which decrease with aperture—even though the wavelength is very short, 8.6 pm for 20 keV electrons, the aperture is so small that diffraction effects are significant). With the high-performance instruments currently available, a resolution of about 5 nm is possible (see, for example, Booker, 1970).

Resolutions obtained in practice are usually much worse than this. As discussed above, if the specimen has insufficient contrast a very long exposure time will be needed during which the instrument may be affected by vibration, stray magnetic fields or drift in the accelerating voltage or lens currents. Moreover, the effects of the electron beam are not confined to the specimen surface. The primary beam electrons will make several collisions inside the material before losing their energy, so the beam will in effect spread out below the surface (figure 2.18). This spreading will increase with the primary electron energy, and decrease with increasing atomic number of the solid; for 15 keV electrons the spread can be as large as 0.25 μm in gold or 1.5 μm in aluminium. The volume of specimen which interacts with the primary beam is therefore of the order of a 1 μm cube and it may include features (for example precipitates) having quite different emission characteristics from those at the surface. The backscattered primary electrons will have equivalent penetrating power to the original beam and will thus arrive at the collector

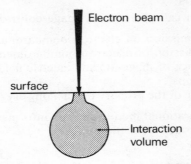

Electron beam

surface

Interaction
volume

Figure 2.18. Schematic diagram of the spreading of the electron beam below the specimen surface, caused by random collisions with atoms of the material. The exact size and shape of the interaction volume depends on the energy of the beam and the mean atomic number of the material

from an area roughly 0.5 μm across even if the original electron probe is vanishingly small. This effect clearly limits the resolution quite drastically. However, secondary electrons have much lower energies ($\leqslant 50$ eV) and can only penetrate 5–10 nm of solid. The resolution obtainable is much greater but it is evident that there is little point in improving the design of instruments intended for surface studies; the resolution is limited in practice by the electron scattering within the specimen rather than by the instrumental resolution of around 5 nm.

If the specimen is very thin, spreading of the beam cannot occur and it is worth using a finer electron-probe. The scanning microscope can then be conveniently used in transmission (with the collector placed below the specimen) and resolutions of 0.3 nm have been achieved by this technique (the 'Crewe STEM' described in chapter 8). The photomultiplier also acts as an image intensifier in this arrangement, so that thicker specimens can be used than in the normal transmission microscope.

As has been hinted in the preceding sections, emission processes and contrast mechanisms other than primary and secondary electron emission can be used in the SEM. It is important always to remember that the beam will interact with quite a large volume of the specimen and that the resolution in each case will depend upon the proportion of the emission from the interaction volume that actually arrives at the detector.

2.5 ELECTRON BEAM/SPECIMEN INTERACTIONS: CONTRAST MECHANISMS

The discussion of the mechanism of image formation in earlier sections started from the fact that when a solid is bombarded with an electron beam, electrons of various energies were emitted. It was also mentioned that other effects could be exploited to extract extra information. It is clearly essential to understand these interactions more fully in order correctly to interpret the images, and to realise the full potentialities of the microscope. In this section we shall discuss the following effects, explaining how each can give rise to image contrast

(1)   Electron emission on electron bombardment
(2)   Photon emission on electron bombardment
(3)   The influence of magnetic and electric fields on the trajectories of secondary electrons
(4)   Channelling of the primary beam along crystal planes in solids

The consequent operational modes and the information obtainable from each will be summarised in section 2.5.8.

2.5.1   Electron Emission

The spreading of the electron beam within the specimen is caused by the multiple collisions suffered by the primary electrons while losing their energy. Each collision of sufficient energy can, in principle, result in the emission of an electron from the specimen and figure 2.19 shows,

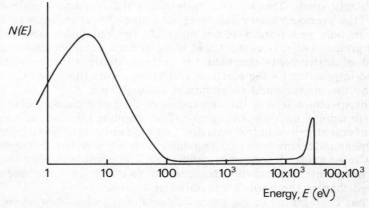

Figure 2.19. Schematic diagram of the energy distribution of secondary electrons emitted on bombardment of a specimen by high-energy (30 keV in this case) primary electrons

schematically, the distribution of energies among the emitted electrons. Some of the collisions are elastic and in such cases the primary electrons may leave the specimen with no significant energy loss. The majority are, however, inelastic and there is a pronounced peak in the distribution of emitted electrons at energies between 0 and 50 eV.

The shape of the secondary electron spectrum varies little from material to material. The peak is shifted to slightly higher energies for solids with lower work functions, though the total secondary emission is greater for materials with high work functions. A little contrast can therefore be expected, resulting from differences in work function in different regions of the specimen. The amount of emission in the range 100 eV–primary energy, on the other hand, increases linearly and quite rapidly with atomic number up to about $Z = 45$, then more slowly for the higher elements. If the secondary electrons are excluded by biasing the collector grid negatively (section 2.3.5) some contrast due to variation in atomic number can be expected even with a smooth specimen that shows no other contrast. Atomic number contrast should under optimum conditions be sufficient to distinguish elements with adjacent atomic numbers up to $Z \approx 20$.

If the surface of the specimen is not very smooth the contrast due to the surface relief ('topographic' contrast) is likely to swamp the atomic-number contrast. This is because the yield of primary and secondary electrons is strongly dependent upon both the angle of incidence of the electron beam and the angle of collection of emitted electrons. The yield improves as the electron beam and angle of collection become more nearly parallel to and normal to the surface, respectively. The angle between beam and collection direction is fixed at about 90°, so maximum yield occurs where the beam is parallel to the surface: unfortunately the image is then completely foreshortened, and an angle of 30–45° between the beam and the specimen normal is usually chosen. Clearly, variations in the local angle of inclination of the surface cause variations in the intensity of the collected electrons. Moreover, the geometry of the surface may prevent some of the electrons from reaching the collector, as illustrated in figure 2.20. This effect is even more pronounced for primary

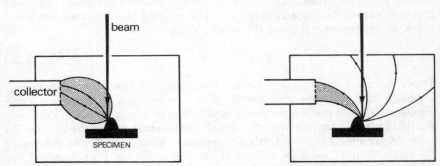

Figure 2.20. The enhancement of topographic contrast in the scanning electron microscope through differences in the efficiency of collection of secondary electrons, depending on their initial trajectories and the local topography; surfaces facing the collector appear brighter than those facing away from it

electrons; secondaries with a wide variety of initial trajectories are attracted into the collector, but the trajectories of the primaries, with energies around 30 keV, are little affected by a collector potential of 250 V. For the same reason of course, the efficiency of collection of primaries is much lower and secondary electrons contribute far more to the image. Topographic contrast can easily exceed 100 per cent.

The energy-distribution plot in figure 2.19 has been drawn as a smooth curve but it does contain some very useful fine structure. When an electron in an inner shell of an atom in the specimen is ejected by a high-energy electron (from the primary beam), the atom can return to its ground state or to some other lower energy state, in two ways. In each case an outer-shell electron drops down into the inner-shell vacancy and energy must be released. The energy release is accomplished in one case by photon emission and in the other by emission of another outer-shell electron. The photon emission results in the characteristic X-ray spectrum, while the electron emission is known as the Auger effect after its discoverer (Auger, 1925; for comprehensive reviews see Bergström and Nordling, 1965, Siegbahn *et al.*, 1967 and Chang, 1971). These effects are important in microstructural analysis because the energies of the photons or electrons that are emitted are characteristic of the elements from which they come (and do not depend on such factors as the primary beam energy). Measurement of these energies therefore gives a qualitative chemical analysis—on an extremely fine scale—and measurement of the intensities of the emission gives a quantitative analysis the accuracy of which will depend upon how well the processes are understood.

The possibilities of compositional analysis using the X-ray emission are discussed in section 2.5.3 and in chapter 3. For the Auger effect, the energy of the Auger electron outside the atom, E_A, is found as follows. The primary beam ejects an electron from an inner shell x, leaving the atom with energy E_x. An electron in an outer shell y drops down to the x shell and the consequent energy release is effected by the emission of the 'Auger electron' from, say, an outer shell z, leaving the atom with two electron vacancies and an energy E_{yz}. The emitted electron requires some work E_w to remove it from the atom. Then

$$E_A = E_x - E_{yz} - E_w \qquad (2.20)$$

These energies will clearly be characteristic of the atom concerned, although the equation above will give a maximum value since once outside the atom the Auger electron also has to escape from the crystal; E_w will increase rapidly with the depth of the atom inside the crystal. In fact, Auger electrons used in microanalysis have energies in the range 50–1000 eV and can only escape from atoms within 1 or 2 nm of the surface. The effective interaction volume is therefore very small and in principle the technique should be able to give quantitative analysis with high spatial resolution—perhaps 20 nm. This has not yet been achieved. The main problem is that the Auger electron current is about 1000 times weaker than the usual secondary electron current; the Auger signal is extracted by measuring the velocity (hence energy) of the electrons and differentiating this signal electronically to show the Auger peaks. High primary-beam currents must be used to obtain sufficient signal and thus the highest resolution cannot be achieved. High-brightness guns are very useful for Auger spectroscopy and field-emission sources are quite convenient to use; the work must be carried out in an ultra-high vacuum in any case, to avoid mere analysis of the surface contamination. Auger spectroscopy and its applications are discussed more fully in chapter 3.

2.5.2 Light Emission

The emission of visible light on electron bombardment is called cathodoluminescence; it is a less general effect than electron emission. The best known cathodoluminescent materials are zinc sulphide and the other phosphors used in cathode-ray tubes although many semiconductors, minerals and biological materials show the effect. The intensity and wavelength of the light depend on the material, so some qualitative chemical analysis may be performed by measuring these parameters.

The principle of image formation is that the emitted photons are collected, converted into an amplified electrical signal, and used (instead of the emitted electron signal) to modulate the display CRT. An example of a mineral image obtained first by using electron emission and then by using photon emission is shown in figure 2.21.

The collection of photons is straightforward—the scintillator crystal is simply omitted from the light pipe. A lens may be added to collect photons over a wider angle. The photons pass directly into the light pipe, enter the photomultiplier and are then converted into an electrical signal and amplified. Photons cannot, of course, be attracted into the collector by electric fields and the efficiency of collection is thus rather low. Another disadvantage is that delay times for cathodoluminescence (unlike electron emission) can be very long; the high-persistence phosphors used on the display screen of the SEM itself are extreme examples. For good resolution the scan rate must consequently be very low, otherwise adjacent parts of the image will become confused. The resolution also depends on the optical properties of the material. If it is optically transparent then photons from the whole of the interaction volume (section 2.4.3) will reach the collector, impairing the resolution. In general, one cannot expect a resolution better than 100 nm using this method, but this (and the depth of field available) is still an improvement on the optical microscope.

2.5.3 X-ray Emission

The idea of chemical analysis employing X-ray emission from a specimen bombarded with electrons was introduced in section 2.5.1. A typical spectrum produced in this way is shown in figure 2.22. The 'characteristic' lines, K_α, K_β, L_α, and so on are produced when electrons ejected from the K, L and other atomic shells in the target element are replaced by electrons from a higher-energy shell. The decrease in energy of the electron falling into a lower energy level is compensated by the emission of a photon of appropriate frequency, ν. The energy decrease and hence the photon frequency is characteristic of the element concerned, as expressed in Moseley's law

$$\sqrt{\nu} \propto Z - C \qquad (2.21)$$

where Z is the atomic number of the target element and C a constant for each type of characteristic line (K_α, L_α, etc). Measurement of the

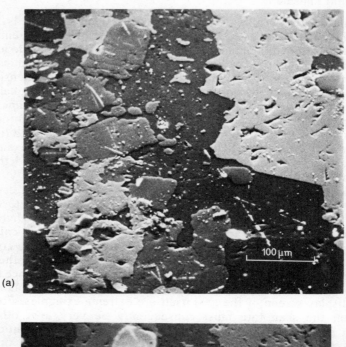

(a)

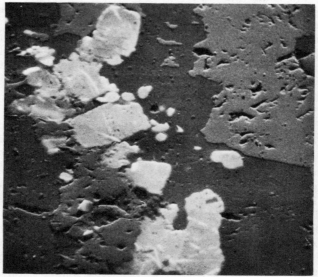

(b)

Figure 2.21. A comparison of images of a mineral specimen made with (a) normal emissive mode of operation, (b) cathodoluminescent mode. (Courtesy of Cambridge Scientific Instruments Ltd)

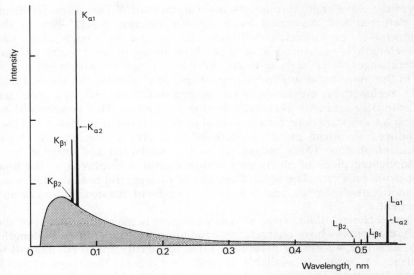

Figure 2.22. X-ray emission spectrum for a molybdenum target showing the K and L characteristic lines (approximately to scale) and, schematically, the continuous spectrum (whose shape will vary according to the conditions)

wavelength and intensity of the photon emission, which lies in the X-ray band, gives qualitative and at least semi-quantitative chemical analysis.

In the SEM this analysis can be carried out on a very small scale. The electron beam can be held stationary on a point of interest, and the X-rays emitted can be analysed. If the beam is scanned as usual and the intensity of the X-ray emission that is characteristic of a particular element is used to modulate the display CRT, a picture of the distribution of that element in the sample will result. This is most useful in metallurgical or mineralogical work as discussed in chapter 3.

The efficiency of X-ray production is low. The parameter f in section 2.4.2 may be only 0.001 so, for comparable image quality, X-ray images may require exposures 10^3 times greater than electron images. This is a serious limitation and X-ray images are usually very noisy.

The penetrative power of X-rays is very great, so photons from the whole of the interaction volume contribute to the image and the resolution is limited to about 1 μm. The collection efficiency is low (X-rays cannot even be focused with a lens) so scan rates must be low and counting times long for quantitative work.

Standard SEMs are not fitted with X-ray detectors (spectrometers), but two different types are available as accessories. In the first, a crystal of known interplanar spacing, d, is used. The wavelength, λ, of the X-rays is measured by diffraction from these crystal planes; when diffraction occurs, the Bragg law

$$\lambda = 2d \sin \theta \tag{2.22}$$

gives the wavelength through the measurement of θ. The diffracted beam is detected and measured by a counter (section 3.3.3). This is the 'dispersive' or 'wavelength-dispersive' method in which only one wavelength is measured at a time. It is usual to have eight or nine interchangeable crystals to cover the whole range of wavelengths.

In the 'non-dispersive' or 'energy-dispersive' method the whole range of wavelengths is measured and displayed simultaneously as a spectrum of intensity against wavelength or atomic number. This is achieved by using a solid-state detector (described in more detail in section 3.3.2) that produces a voltage pulse proportional in energy to the energy of the incident photon. These pulses are sorted, measured and counted by a complicated piece of electronics whose output is displayed as the final spectrum on a CRT. The signal from any of the spectral peaks can be used to modulate the image, and thus give a picture of the distribution of any element.

The technical difficulty in all X-ray analysis is the measurement of the light elements, whose characteristic radiation has such long wavelengths (for example, 4.5 nm for carbon K_α) that it is easily absorbed, difficult to diffract by ordinarily available crystals, and outside the scope of solid-state detectors. The wavelength-dispersive method is currently superior in the analysis of light elements, and in accuracy and resolution for most of the other elements. However, these advantages are not conclusive when a spectrometer is used as an accessory on an SEM; X-ray microanalysis can only be semi-quantitative in this instrument, because special constraints are imposed upon the electron and X-ray optics and on the specimen if accurate analysis is required, as will be discussed in the next chapter. The very much higher speed of analysis of energy-dispersive spectrometers, allied with an adequate performance for many applications, makes them much the more suitable choice for use with the SEM.

Chapter 3 is devoted entirely to the theory and applications of microanalytical techniques, which are both important scientifically and very widely used in industry.

2.5.4 Electrical Effects Induced in the Specimen

The electrical effects induced by the electron beam in solids are most important, especially in the study of semiconductor materials and solid-state devices. The energy of electrons in the primary beam is, say, 30 keV, and the energy required for impact ionisation in solids varies from a few electron volts to tens of electron volts. We should therefore expect one primary electron to generate some thousand excess free electrons and an equivalent number of ions ('holes'). It is found experimentally that on average between 0.5 and 1 electron is emitted from the surface per incident primary electron, and the complementary number flows to earth as the specimen-to-earth current. The vast majority recombine. The time for recombination is only about 10^{-12} s in metals, which is far too short for these electrons to be exploited in any way, and in any case there is no shortage of free electrons in metals. In semiconductors on the other hand, the recombination (relaxation) time varies with purity and may be as much as a few seconds. These excess current carriers (both electrons and holes) will have a large effect on the limited conductivity of such materials. Also, the carriers generated in one point will diffuse towards regions of lower carrier concentration, obeying normal diffusion laws, and voltages will be established wherever the carriers encounter regions of different chemical composition (for example where impurity atoms have clustered around a dislocation, causing local variations in the internal electric field in the material).

Two effects are thus detectable if the recombination time of carriers is not too short: (1) the enhancement of conductivity, by an amount controlled by local impurity levels, at each point struck by the beam; (2) the establishment of voltages at inhomogeneities in the specimen. The detection of surface voltages is treated in the next section. The conductivity effect can be monitored by applying a potential difference across the specimen from an external battery, and using the magnitude of the resulting current to modulate the brightness of the display CRT. An image of the conductivity variation in the specimen can thus be formed.

As with cathodoluminescence, the decay time may be so long as to necessitate very slow scan speeds, and as with X-ray analysis, the whole interaction volume will contribute to the image so that the resolution will be of the order of 1 μm. Sideways diffusion of the carriers will also contribute to a resolution loss which will be more pronounced the longer the decay time and the higher the diffusion rate.

2.5.5 Voltage and Field Contrast

If different areas of the specimen are at different potentials, due to the effects discussed in the last section, to the accumulation of charge on an insulating specimen, or to the application of an external e.m.f., the production of emitted electrons is unaffected but their collection is modified in two ways. First, the long-range field between the collector and the specimen will vary from point to point with the surface potential on the specimen, and the efficiency of collection will likewise vary. In the extreme case, if the specimen and collector potentials are equal no secondary electrons will arrive at the detector, and the contrast (the change in signal divided by the original signal) is 100 per cent. However, it would be wrong to assume that the contrast is linear with the ratio of specimen potential to collector potential, because it is strongly affected by the precise geometry of the specimen and collector. Since the essential effect is the perturbation of the trajectories of the secondary electrons, higher contrast can be achieved by limiting the collector aperture, as was shown in a detailed investigation by Everhart et al. (1959). It was found that a simple slot aperture sufficed to improve the contrast to about 5 per cent per volt, although the exact position of the aperture was important. Contrasts below 5 per cent are scarcely visible so that a potential difference of one volt is the smallest that can conveniently be seen.

Smaller potentials (say down to 0.1 V) can of course be seen by limiting the aperture still further if resolution is not important or if the scan time can be proportionately increased to overcome the noise.

Secondly, the trajectories of the secondary electrons are also affected by electric fields on or near the specimen surface, as can be produced during the operation of semiconductor devices, and again contrast is obtained. Although at first sight these two mechanisms appear similar, since the fields in any space are determined by the potentials on surrounding surfaces, it is convenient to separate their effects on the contrast; the effect of surface fields is limited to a region near the specimen surface, whereas the surface potentials affect the field in the whole region between specimen and collector. For example, the potential difference between two adjacent regions on a semiconductor specimen can be too small to give voltage contrast between the regions; but if the boundary separating them is very narrow the surface field will be intense and will be visible as a streak on the image. It is perhaps helpful to consider voltage contrast as arising from the field in the direction joining specimen and collector, and surface field contrast as arising from fields transverse to this direction.

Oatley (1969) has also developed an ingenious method for distinguishing between voltage (or field) contrast and contrast due to surface topography (or any other mechanism) as long as the voltages or fields are produced by an externally applied e.m.f. The external e.m.f. is not applied continuously but pulsed in a square wave, and the primary beam is chopped at double the frequency of the specimen bias. The result is that for exactly half of the primary beam pulses a bias voltage is being applied to the specimen, and the collector receives a signal containing all the contrast, while for the other half of the pulses the bias is off and the signal contains all contrast except that due to voltages and fields on the specimen. The two signals can be electronically sorted and subtracted to form an image showing only voltage and field contrast.

It is quite easy to observe surface potentials or fields in the SEM but much harder to make quantitative measurements. This is partly because of the geometric effects mentioned earlier but also because electron trajectories are affected by both potentials and fields, which of course always coexist. Energy analysis of the secondary electrons does help, because the energy of a secondary electron arriving at the collector depends among other things on the collector-specimen potential difference, but even this is modified by the presence of surface fields. The problem can be solved, however, by using a more complicated detector, which has extra electrodes carrying potentials that can be varied from outside the vacuum system (Banbury and Nixon, 1969, 1970). Different values of these potentials optimise the detector for different functions, such as voltage contrast or topographic contrast. A linear sensitivity of between 8 and 13 per cent per volt (depending on the range required) is given by this new detector, and surface fields (transverse fields) have very little influence on this contrast. In standard instruments, on the other hand, only the qualitative features of the contrast can be employed;

nevertheless the method is invaluable for studying the operation of microcircuits, as discussed in section 2.6.3.

2.5.6 Magnetic Field Contrast

Ferromagnetic domains can be observed quite well in the scanning electron microscope by means of two distinct contrast mechanisms

(1) The magnetic fields outside the specimen, which run between adjacent domains (leakage fields) are strong enough in some materials to affect the trajectories of secondary electrons; the consequent alteration of collection efficiency gives rise to contrast as with surface electric fields, mentioned in the last section. Images of the surface domain structure are obtained, which again can be improved by limiting the aperture of the collector (Booker, 1970; Joy and Jakubovics, 1968), or, more effectively, by employing the Banbury–Nixon detector (section 2.5.5).

(2) The trajectories of electrons scattered inside the specimen are also modified if the specimen is magnetic. Only the change of trajectory in the direction of the collector is important, so only the magnetisation component that is perpendicular to the specimen-collector line gives rise to contrast (an approximate measure of the magnetisation direction can therefore be obtained). This contrast is only a few per cent and appears only if the specimen surface is tilted away from the plane normal to the electron beam. However, high resolution is often unnecessary when studying domains, so the beam diameter and current can be increased to reduce the noise and permit electronic enhancement of the contrast (Philibert and Tixier, 1969; Fathers et al., 1973; Newbury and Yakowitz, 1973). An example of magnetic domains imaged with this type of contrast is shown in figure 2.23.

These methods can profitably be used to investigate domain structures and their distortions and rearrangements under magnetic field or mechanical stress, but magnetic contrast can also blur other features of interest. An example is the very fine-grained structures formed when tungsten carbide is sintered with cobalt; SEM images of this material are greatly improved by demagnetising the specimen (Lardner and Boucher, 1972). Demagnetisation with an a.c. coil does not remove the magnetisation but creates equal volumes of oppositely magnetised domains, and makes the domains grow to attain a lower energy configuration. The larger domains have a smaller perturbing effect on the secondary electron trajectories.

It must be remembered when performing SEM experiments on magnetic materials that the final lens is itself a strong magnet. Although it is well shielded, the field at the specimen surface is usually in the range 1–20 mT (that is, 50–1000 times stronger than the earth's field) and must be taken account of in experiments with weak magnetic fields.

2.5.7 Electron Channelling and Crystallographic Contrast

This contrast mechanism, unlike all those discussed so far, is only applicable to crystalline specimens, for which it gives information about

(a)

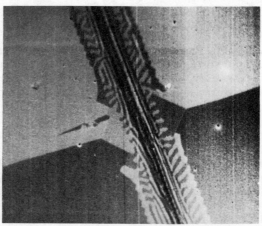

(b)

Figure 2.23. Magnetic domains in iron–silicon alloy (as used for transformer cores) taken on a standard scanning electron microscope. (a) $3\frac{1}{2}$% Si, (110) textured. The domains are 180° type, about 5 μm wide. Angle of incidence 65°, tilt axis horizontal. (b) $4\frac{1}{4}$% Si, (100) textured. Field of view approximately 350 μm across. The complex pattern of 180° and 90° domains is induced by interaction between the magnetostrictive properties of the material and the stresses caused by the scratch on the surface. Angle of incidence 45°, tilt axis horizontal. (Courtesy of D. J. Fathers)

the crystal symmetry, unit cell size and crystal orientation. No diffraction pattern can be obtained in the SEM since the scattered electrons have almost all suffered inelastic scattering. However, the incident beam experiences a diffraction effect when it enters the crystal, known as electron channelling. As the name implies, the electrons are channelled between crystallographic planes, the amount of channelling per plane depending on its packing and its spacing. This will cause a variation in the secondary electron emission with the angle of incidence of the primary beam because of the different primary electron distribution beneath the surface that exists when the beam is channelled along different planes. Corresponding contrast will be found in the image (superimposed upon other contrast present) since the angle of incidence of the beam varies during the scan.

This is the weakest effect that can be used in image formation, and it is not observed in normal use. The contrast is easily masked by other variations; the angular range of the beam is very small at high magnifications; and the beam may cross several grains during its scan (this will give some 'orientation contrast' between grains). If a large single crystal is used at low magnification then these electron channelling patterns (also called Coates patterns after the discoverer) can be observed (figure 2.24)

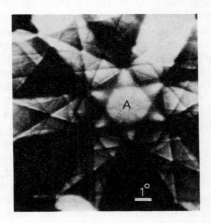

Figure 2.24. Channelling patterns obtained from a single crystal of silicon, taken at 20 kV in an unmodified microscope. The threefold symmetry about point A identifies it as a [111] direction; the centre of the picture corresponds to normal electron beam incidence, so the orientation of the crystal surface is about 3° away from (111). (Courtesy of C. G. van Essen)

but the position of the beam naturally changes as the angle changes and different parts of the angular information are obtained from different parts of the specimen. A better method is to abandon the normal practice of scanning the beam over a raster, and instead to rock it through as large an angle as possible while it remains directed at one point on the specimen. One way of doing this is shown in figure 2.10b (van Essen *et al.*, 1970). This should be compared with figure 2.10a, when it will be seen that only the upper scan coils are used, and the lenses are used at different strengths (the upper scan coil is effectively placed at the object point of the final lens). The beam divergence should be low and the beam as intense as possible since the sharpness of the patterns depends not on the spot size but on the beam divergence.

Using these techniques, channelling patterns have been obtained from accurately selected areas as small as 1 μm diameter and it should eventually be possible to reduce this by at least one order of magnitude. Two other techniques are available for measuring the orientation of selected areas of bulk specimens. First, the method of microbeam X-ray diffraction. With great care and skill, diffraction patterns from regions 5–10 μm diameter can be obtained, and with even more skill they can be correlated with features in the microstructure. The channelling pattern method is obviously more convenient and accurate for orientation determination, although the microbeam X-ray diffraction method gives other information as well (for a review see Hirsch, 1960). The second is the technique of measuring the Kossel line pattern produced when a divergent X-ray beam is transmitted through a crystal (see for example Yakowitz, 1964). The characteristic X-rays are strongly diffracted when they satisfy the Bragg angle for various crystallographic planes and, with a divergent incident beam, a series of curved lines, related to the crystal orientation, is produced on a photographic plate placed in the transmitted beam. In the scanning electron microscope the specimen can be used as both target for the generation of X-rays and as specimen, since characteristic X-rays are generated by the primary beam and escape from the specimen by transmission through its top layer. A photographic plate or film placed above the specimen (and containing a hole to admit the primary beam) will then record the Kossel pattern. A comparison of the two techniques (channelling and Kossel patterns) is given by Joy *et al.* (1971); they have similar accuracies, the channelling pattern method is more convenient to use since the patterns can be displayed on a viewing screen but the Kossel line technique requires no special apparatus (provided the microscope has a sufficiently high accelerating voltage to stimulate the required characteristic line). Channelling-pattern information comes from the first 50 nm of the specimen whereas that for Kossel lines is taken from a depth up to 100 times as great.

Selected-area channelling patterns are most readily identified by comparison with a map of all possible orientations (figure 2.25, for copper at 30 kV), when the orientation of an unknown sample can be determined to better than 0.1°. The spacings of the bands on the map are dependent upon accelerating voltage and lattice parameter, and the patterns are distorted by magnetic fields and blurred by high densities of lattice defects. The map also clearly reflects the crystal symmetry. There is a close relationship between ECPs in the scanning microscope and Kikuchi patterns in the transmission electron microscope (chapter 5); a detailed discussion of the contrast is given in Hirsch and Humphreys (1970).

Channelling patterns are better observed by using the current flowing to earth through the specimen, rather than the emitted current, to modulate the beam. The two currents are almost complementary, but the specimen current is not affected by the efficiency of collection of secondaries (figure 2.20) and hence is less sensitive to surface topography. Since topographic contrast is usually much stronger than crystallographic contrast, this is very useful when a normal scan is used. It also has advantages for the

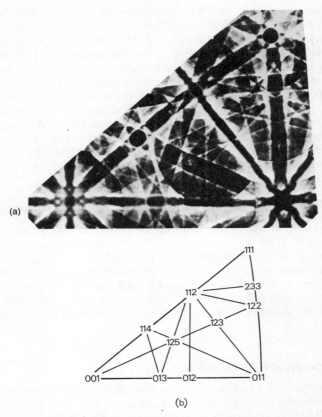

Figure 2.25. (a) Channelling pattern map for copper, taken in the absorptive mode at 30 kV. This is a composite photograph made from several exposures on a spherical single crystal and may be used to identify small segments of the map as are found in selected-area channelling patterns. (b) A diagram showing the main crystallographic directions appearing on the map, which covers the standard stereographic triangle. (Courtesy of C. G. van Essen)

selected-area method; although the beam is then directed at the same spot all the time, it has a finite width and may span a feature of the surface relief.

2.5.8 Operational Modes

All the major methods of obtaining image contrast have now been discussed with the exception of contrast in the transmission mode (briefly mentioned in section 2.4.3) since contrast theory in this mode is essentially identical with that for the direct-transmission electron microscope, treated in chapter 5. Since there is no image-forming lens in the SEM, it is easier to achieve the highest resolutions in this way rather than in the transmission microscope, and it is much easier to 'process' the image, for example it may be formed from electrons of a selected energy, or the contrast may be electronically enhanced. The disadvantages of 'scanning transmission' lie in the noise problem and the greater cost of an instrument comparable with the best transmission microscopes. Scanning-transmission microscopes will be discussed more fully in chapter 8.

At this stage we can usefully summarise the various operational modes of the SEM together with the types of information that can be obtained from each (table 2.3). It will be evident by now that the mode selected will depend entirely on the information sought, and that the specimen itself has a profound influence upon the information possible, the operational mode and, in particular, the resolution obtainable.

2.6 SOME APPLICATIONS OF SCANNING ELECTRON MICROSCOPY

Since the method of use of the scanning electron microscope is so closely linked with the type of specimen and with the information which is sought, several of its applications have already been discussed. In the following sections some examples are given of the types of problem in materials science that are most often investigated on the SEM, involving the study of surface relief, chemical composition, semiconductor materials and solid-state devices; finally, dynamic experiments are considered in section 2.6.4.

2.6.1 The Study of Surface Relief

The investigation of surface relief, or topography, is the most common task of the SEM (for example, figure 2.1). It is the high depth of field of the SEM, rather than its sensitivity to surface relief, that makes it so valuable for topographic studies. Figure 2.26 shows a fracture surface in a pure iron specimen, in which the areas of intergranular fracture and cleavage are clearly distinguished. This photograph could not have been taken on a conventional optical microscope nor on a transmission electron microscope, and 'fractography' as a method of studying brittle, ductile or fatigue failure has made great progress. The annual scanning electron microscope symposia held in Chicago usually contain several papers on fractography. Particularly worthy of mention are: those by Kanai and Uchibori (1969) who examined the relationship between microstructure and both brittle and ductile fracture in a range of steels, and in their paper published a series of representative micrographs of the different types of fracture; Waldron *et al.* (1970) and Pelloux *et al.* (1970) who have shown that scanning electron microscopy can reveal the important features of fatigue fracture (the crystallographic stage I cracks, structure-sensitive and structure-insensitive stage II cracks, fatigue striations, etc.) without the problems associated with replication techniques used in conjunction with a transmission electron microscope, even though the resolution of the latter is higher; and Scully (1970) who has discussed the application of the

Table 2.3. *Information available from various operational modes*

Mode	Modulation effect	Modulation device	Best resolution	Information available
Emissive	Secondary electrons	Electron collector	5–15 nm	Topographic Work function Surface potential and field (hence electronic device operation) Magnetic field Chemical analysis of surface layer using Auger electrons
Reflective	Backscattered primary electrons	Electron collector	50–250 nm	Topographic Atomic number
Transmitted	Transmitted primary electrons	Electron collector (beneath specimen)	0.5 nm	As for transmission electron microscopy; image processing possible
Absorptive (Specimen current)	Specimen-to-earth current	Specimen current amplifier	5–15 nm	As for emissive mode, but topographic contrast partially suppressed (also used for high beam current or if normal detector obscured by specimen)
Conductive	Excess carriers generated by primary beam	Current through specimen from external source is amplified	>1 μm	Conductivity and impurity variations in semiconductors
Cathodoluminescent	Photons in visible band	Photomultiplier	>1 μm	Qualitative chemical analysis for some materials
X-ray	Photons in X-ray band	X-ray spectrometer (Emissive or absorptive mode)	>1 μm	Semi-quantitative chemical analysis for $Z \geqslant 10$
Crystallographic	Angular variation of electron channelling		0.05° from 1 μm selected area	Crystal orientation Lattice parameter Qualitative information on defect density and magnetic fields

Figure 2.26. Fracture surface of pure iron specimen, broken at 77 K. The large grain filling most of the field of view has fractured by cleavage and the fine white lines are 'river markings' typical of this failure mode; the smoother but more undulating areas above this grain are typical of intergranular fracture

scanning electron microscope to the study of stress-corrosion cracking, in which the specimens are so rough, but contain such significant detail, that the SEM is virtually the only instrument suitable for their examination. The technique of selected-area channelling patterns (section 2.5.7) will probably become important for fractography, since it is often necessary (especially for stress-corrosion fracture) to determine the crystallographic plane of fracture in order to discover the mechanism.

The high depth of field also enables one to obtain three-dimensional information very rapidly and accurately. One new metallurgical technique that has been developed because of this facility, is that of deep etching of a multiphase alloy. Figure 2.27 shows the eutectic phase in an aluminium–silicon alloy which has been deep-etched with dilute HCl. The aluminium-rich phase has been removed and the continuous network of silicon crystals is easily seen. Before this technique became possible, such information was obtained by taking a series of optical micrographs, grinding away a thin layer of the specimen between each one, and reconstructing the three-dimensional structure—a heartbreakingly laborious method. For a discussion of the technique required for accurate three-dimensional measurement the chapter by Lane in Hearle et al. (1972) should be consulted, and examples of applications of deep-etching to the study of eutectic phases can be found in Day and Hellawell (1967) for Al–Si and Day (1969) for cast iron.

Problems associated with fibres and polymers are often very conveniently studied on the SEM, partly again because of the high depth of field, and partly because the resolution possible on an optical microscope is

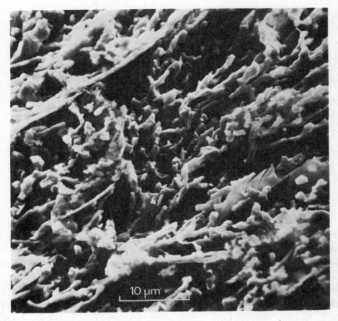

Figure 2.27. Aluminium–silicon eutectic, deep-etched with hydrochloric acid to dissolve the Al-rich phase and reveal the branched structure of the Si-phase

often inadequate for these problems. Fractography is again very useful, and figure 2.28 shows a nylon fibre that has been split to reveal internal defects, in this case titanium dioxide inclusions. Correlation of such information with the parameters of the production process can clearly lead to the production of better fibres. Straining stages inside the specimen chamber can also be used to examine fibres or complete fabrics at any stage prior to actual fracture, as discussed in section 2.6.4. However, damage in the electron beam can sometimes be a problem with these materials. A related class of materials is that of natural products such as cotton, wool and wood. Scurfield and Silva (1969) have made a particular study of wood and have shown that the high resolution and high depth of field of the SEM, plus its wide range of magnification, can again be used to advantage to complement other techniques.

The examples so far have been of relatively large-scale topographic effects and of macroscopically rough structures. The SEM is not so good at revealing small surface relief, such as slip lines on an otherwise smooth surface. Figure 2.29 shows slip lines on the surface of a niobium single crystal in both the SEM and the optical microscope (using the Nomarski interference contrast technique) and in this case the SEM is inferior; the minimum step-height detectable is probably about 10 nm, compared with about 2 nm for optical interference contrast and 0.5 nm for the best multiple-beam or holographic interferometry methods. The contrast is

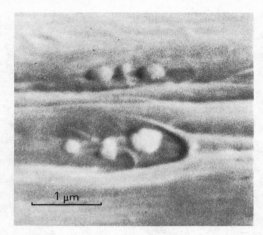

Figure 2.28. Nylon fibre, split and peeled to show internal defects. The small inclusions are of titanium dioxide. Note the very high magnification of this photograph. (Courtesy of W. D. Emery, Unilever Research Laboratory)

higher for sharp steps such as cleavage steps (rather than slip lines) and is somewhat enhanced by the use of 'mirror microscopy' in which the specimen is held at about the same potential as the cathode. The primary beam does not then penetrate the specimen but is reflected. Rapid changes in the local electric field occur at steps on the surface and image contrast is produced from these features. The contrast can also be improved by the use of a special detector (Banbury and Nixon, 1969) which is sensitive to the initial trajectories of the secondary electrons (see also section 2.5.5).

2.6.2 The Study of Chemical Composition

Qualitative information can be obtained about chemical composition from atomic number contrast and from cathodoluminescence, although the latter has rather limited application. Figure 2.30 shows a flat, polished specimen of a cast iron in which the very different atomic numbers of iron and carbon have given a high contrast. This is a useful mode of contrast for metallic and ceramic specimens, and also for polymers and biological materials in which heavy element compounds may be present in a matrix of light elements. A good example has already been seen in figure 2.28, since the titanium dioxide particles are emitting more strongly than the nylon matrix. Atomic number contrast is, however, fairly weak and is easily obscured by topographic contrast.

As discussed in sections 2.5.1 and 2.5.3, analysis of the X-ray and Auger electron emission provide much more powerful means of studying chemical composition and its relation with the microstructure. Microanalytical techniques are so important that their theory and application are treated in greater detail in the whole of the next chapter.

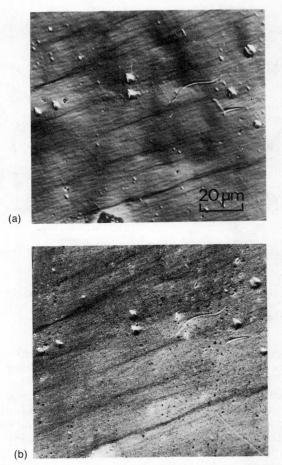

(a)

(b)

Figure 2.29. Comparison of optical interference contrast (Nomarsky method) with SEM for the examination of fine surface relief: (a) Optical interference contrast showing fine slip lines; (b) SEM image of same area with specimen tilted 45° (the foreshortening has been electronically corrected to aid comparison)

2.6.3 The Study of Semiconductors and Solid-state Devices

This field may be divided into three parts: study of the properties of semiconductor materials, investigation of device production techniques, and observation of complete circuits in operation.

The materials themselves may usefully be studied in the SEM in the emissive, absorptive, conductive, cathodoluminescent or crystallographic modes. Semiconductors exhibit so many of the effects observable in the SEM that a vast amount of information is available. Only the X-ray mode is likely to be of limited interest, since the concentrations of the

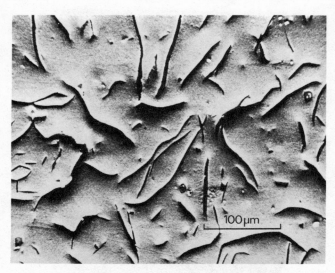

Figure 2.30. An unetched specimen of cast iron containing flaky graphite, which is shown up by atomic number contrast using primary electrons only (the specimen is not completely flat and some topographic contrast can be seen in the matrix around the flakes)

interesting impurities are extremely small and the elements occurring as major or minor constituents in many semiconductors have low atomic numbers.

Integrated circuits are commonly prepared by processes involving deposition of thin films of material onto substrates of another material. Numerous faults occur in this process: low adhesion of the films, holes in the films, cracks or short circuits in the connecting strips, and so on. Figure 2.31 shows examples of the SEM used to study devices in this way, and correlation with the production process and the device performance can lead to marked improvements even in this field, in which there have been such dramatic developments in the last decade. Besides the unique contrast modes of the SEM, many of the faults it can reveal are well beyond the resolution of the optical microscope.

An interesting consequence of the nature of the scanning electron microscope is that it can itself be used for making devices. A fine beam of electrons is available, together with a scanning system that can position the beam anywhere on a specimen with great accuracy. If another deflection coil is added, essentially to switch the beam on and off very rapidly, any desired pattern can be traced on the surface using external controls on the scan coils. If the specimen is coated with 'photoresist', a substance which changes its chemical resistance on exposure to light or electrons, the required pattern can be traced on the specimen, the pattern dissolved in a suitable chemical, and the appropriate film evaporated onto the exposed parts of the surface. This is, in principle, the method used in

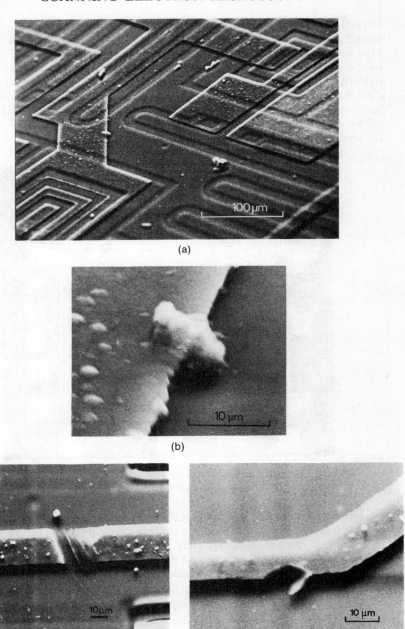

Figure 2.31. (a) An overall view of an integrated circuit in the SEM, plus details of circuits showing various production faults: (b) inclusion at film edge, (c) rupture of film, (d) uneven width of film

normal production processes; in this use the electron beam replaces the light beam and the controls on the scan coils replace the optical mask. However, optical methods are limited in practice by the blurring of the image of a sharp edge caused by diffraction, and much smaller-scale patterns can be produced with an electron beam. Figure 2.32 shows a circuit produced in this way, which enables new, small-scale circuits to be produced quite rapidly for development work (see for example Broers, 1965). Alternatively the SEM may be used to produce the 'master mask' for quantity production (Chang 1971).

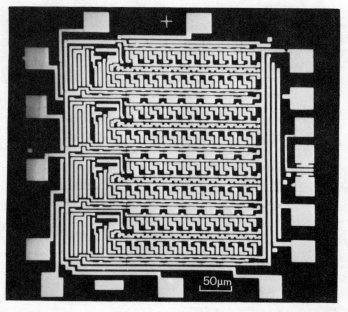

Figure 2.32. A circuit produced by electron-beam processing in a modified SEM. (Courtesy of Cambridge Scientific Instruments Ltd)

The contrast obtained from surface potentials and fields on the specimen provides a unique method for studying circuits in operation. As seen in figure 2.33 the different electrical conditions in the various parts of the circuit can be clearly observed and the effect of changing the electrical (or thermal) conditions can be followed. It is also possible to observe design

Figure 2.33. The use of the SEM in the study of the operation of electronic devices. The circuit is a metal-oxide semiconductor 16-bit parallel-access 4-phase shift register. (a) −30 V on No. 1 pulse line, specimen current amplifier on inner clock pulse line, other connections earthed. (b) Connections to No. 1 line and clock line reversed. The image was produced by mixing the emissive and conductive mode signals, and inactive parts of the circuit can be seen by faint topographic contrast. (Courtesy of Cambridge Scientific Instruments Ltd)

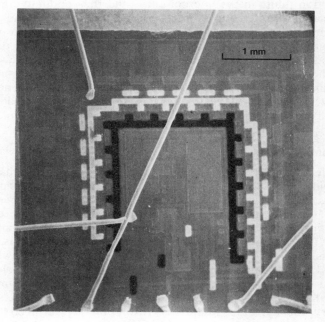

(a)

(b)

or construction faults in the microcircuits, and the conductive mode in particular can give quite detailed information about the electrical properties of the interior of the semiconductor. It is even possible to observe circuits operating in alternating current at frequencies up to the microwave band, by modulating the primary beam. If the beam is flashed on and off at the same frequency as that applied to the circuit, the operation of the circuit will be 'frozen'; this stroboscopic technique has not been very widely used as yet, but clearly has considerable potential (Plows and Nixon, 1968). It may also be used to study mechanical vibration.

2.6.4 Dynamic Studies

Scanning electron microscopes normally have large specimen chambers, often provided with several electrical and mechanical feedthroughs. There is, accordingly, ample opportunity for performing experiments within this chamber and following the process with the microscope. Simple heating and cooling attachments are very easy to design, and have been used (for example) for studying phase transformations. Good progress has been made on the design of micromanipulators (Pawley and Hayes, 1971), that could be used for delicate mechanical or electrical experiments or for micro-dissection of biological material. It must be remembered that images can only be formed at a high rate in the SEM (as is necessary for dynamic studies) if resolution is sacrificed or if a high-brightness gun is available, as was discussed in section 2.4.2. Bearing this limitation in mind, we shall consider some of the possibilities for dynamic experiments.

Several such applications to the study of the manufacture and operation of semiconductor devices were mentioned in the last section, and in most of those cases little or no modification of a standard instrument was needed. Special stages are needed for the study of mechanical properties of materials, and several tensile stages that fit inside the SEM have been designed, for example by Dingley (1969) for metals, composite materials and the like, and Cross et al. (1970) for textile fibres and fabrics. The requirements are rather different for these two classes of materials since fibres and fabrics usually require a much higher extension than metals, but deform at a considerably lower stress. An example of the use of a tensile stage to examine changes occurring in a fabric during deformation is given in the paper by Cross et al. and concerns the extension of a non-woven fabric, formed by bonding fibres together with a resin. Examination under stress showed the way in which the fibres moved and the resin fractured. It should be noted that in such materials the appearance of the specimen under stress might be different from that when the stress is applied and relaxed, since polymers are very sensitive to the rate of deformation; a tensile stage can then give much more meaningful results than examination of specimens that have been deformed some time previously. Examples of applications of the heavier-duty tensile stage to the deformation of metallic alloys are given by Dingley (1970) and Joy and Newbury (1971), the former on superplastic alloys and the latter on pure metals and superplastic alloys to evaluate the use of the selected-area channelling pattern technique for the study of deformation (using the fact that the clarity and regularity of the patterns diminish with increasing plastic deformation of the crystals studied). It is likely that the study of superplastic alloys will be a more productive field for the SEM than will the investigation of 'conventional' plasticity. This is partly because, as mentioned earlier, the SEM is not particularly good at revealing shallow slip steps and partly because it is very good at resolving small lateral shifts, such as occur during superplastic deformation by grain boundary sliding.

The work of Scully in connection with stress-corrosion cracking was mentioned in section 2.6.1 and several workers have used the SEM to observe the results and products of corrosion (see, for example, the chapter by Castle in Hearle *et al.*, 1972). Dynamic studies on corrosion are just beginning. At first sight, a corrosive environment would appear to be far removed from the high-vacuum requirement of an electron microscope, but as long as the corrosive vapours or gases are highly concentrated only in the region of the specimen, the secondary electrons can penetrate them to reach the collector (Thornley, 1960). Lane (1970) has designed an 'environmental control stage' that plugs into the normal specimen carrier and provides a liquid, vapour or gaseous atmosphere around the specimen. The atmosphere is continuously pumped away by the vacuum system and renewed from reservoirs either inside the stage or outside the microscope. It is sufficiently concentrated to provide genuine corrosion but so localised that the operation of the microscope is not impaired. The use of this stage is of course not confined to corrosion studies; the inventor suggests applications as diverse as the investigation of the freeze-drying process and the detection of ionic currents in living nervous systems ('biological microcircuitry').

2.7 IMAGE PROCESSING

Since the image from the SEM is built up one point at a time, and at any instant there is only one electrical signal modulating the display CRT, it is possible to extract extra information by electronically processing this signal. At its simplest, the contrast may be electronically enhanced as in a television set (provided that the signal/noise ratio is high enough). It may also be useful to employ a non-linear amplification so that the highlights (which tend to get burned-out) receive less amplification than the faint regions, or to differentiate the signal so that only sharp changes in electron emission give contrast on the image. Figure 2.34 shows an example of these techniques applied to pictures of tungsten carbide powders. When high topographic contrast is present, much more detail can be seen when the intense highlights are toned down, in this case by differentiating the signal.

In normal operation, the signal from the specimen is used to modulate the intensity of the display CRT. It may instead be used to modulate the direction of the CRT beam, (which is then kept at constant intensity). This

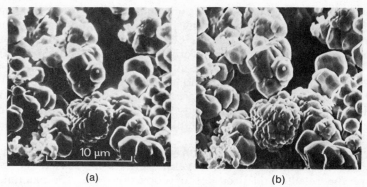

(a)   (b)

Figure 2.34. An example of image processing, using a tungsten carbide specimen in powdered form: (a) standard emissive-mode signal; (b) differentiated signal, which gives contrast only at changes in the signal and hence reveals details in the shadows and highlights. (Courtesy of Wickman Wimet Ltd, Coventry)

method, known as 'Y-modulation' is illustrated in figure 2.35; the cleavage steps can be seen more clearly in the Y-modulated image and a more vivid stereoscopic effect is produced.

Signals from different detectors may also be combined. It is often useful to examine a specimen in two or more modes, and if, say, the signal from the conductive mode is mixed with that from the emissive mode when studying an operating microcircuit (as in figure 2.33), the surface potentials and fields may be more easily correlated with the internal conductivity and the surface topography.

(a)

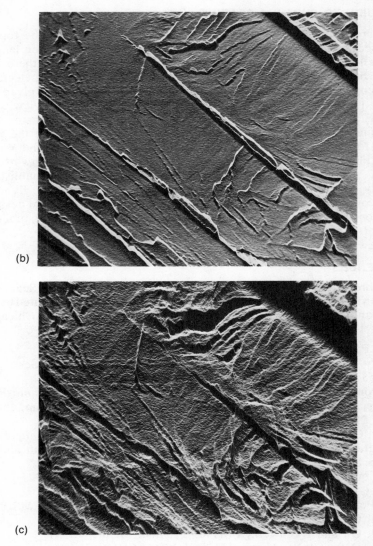

(b)

(c)

Figure 2.35. An example of Y-modulation of the image. Pure iron specimen, fractured at 77 K (as used for figure 2.26). (a) Emissive mode; (b) and (c) Y-modulated images with greater amplitude of modulation in (c)

Finally, the most extreme and versatile example of image processing lies in the use of a digital computer. If a real-time computer is connected between the detector outputs and the visual display an infinite variety of image processing may be carried out: images may be digitally stored, mixed and compared, and quantitative measurements and any desired data processing may be performed.

2.8 GUIDE TO FURTHER READING

There are now several fairly comprehensive references for the topics discussed in this chapter. For electron optics, design of electron lenses and the like, we recommend the book by Hall (1966), though there are several other good books on this subject. Reviews of scanning electron microscopy that are rather more detailed than this chapter are given by Booker (1970), Nixon (1969) and Oatley *et al.* (1965; this review also contains an interesting historical survey). Complete monographs on the technique have been published by Oatley (1972; the inventor of the first successful instrument), Thornton (1968), Hearle *et al.* (1972) and Wells (1974), all of which are very useful. In addition, the *Proceedings of the Annual Scanning Electron Microscope Symposia* have been published annually since 1968 by the I.I.T. Research Institute, Chicago, Illinois. These contain a wealth of original papers on every aspect of SEM design, operation and image interpretation and also include a cumulative bibliography of papers on scanning electron microscopy.

REFERENCES

The abbreviation *SEMS* refers to the proceedings of the annual Scanning Electron Microscope Symposium, for the year given. These are edited by O. Johari and published by the Illinois Institute of Technology, Chicago, Illinois, U.S.A.

Auger, P., (1925), *J. Phys. Radium*, **6**, 205
Banbury, J. R., and Nixon, W. C., (1969), *J. scient. Instrum.*, Ser. 2, **2**, 1055; (1970), *SEMS*, 475
Bergström, I., and Nordling, C., (1965), in *Alpha-, Beta- and Gamma-Ray Spectroscopy*, (ed. K. Siegbahn), North-Holland, Amsterdam, chapter 25
Booker, G. R., (1970), in *Modern Diffraction and Imaging Techniques in Materials Science*, (eds S. Amelincx, R. Gevers, G. Remaut and J. van Landuyt) North-Holland, Amsterdam, 613–53
Broers, A. N., (1965), *Microelectron. and Reliab.*, **4**, 103
Chang, C. C., (1971), *Surf. Sci.*, **25**, 53
Cross, P., Hearle, J. W. S., Lomas, B., and Sparrow, J., (1970), *SEMS*, 81
Day, M. G., (1969), *J. Metals*, **21**, 31
Day, M. G., and Hellawell, A., (1967), *J. Inst. Metals*, **95**, 377
Dingley, D. J., (1969), *Micron*, **1**, 206; (1970), *SEMS*, 329
Everhart, T. E., and Thornley, R. F. M., (1960), *J. scient. Instrum.*, **37**, 246
Everhart, T. E., Wells, O. C., and Oatley, C. W., (1959), *J. Electron. Control*, **7**, 97
Fathers, D. J., Jakubovics, J. P., Joy, D. C., Newbury, D. E., and Yakowitz, H., (1973), *Phys. Stat. Sol. A*, **20**, 535
Hall, C. E., (1966), *Introduction to Electron Microscopy*, McGraw-Hill, New York
Hearle, J. W. S., Sparrow, J. T., and Cross, P. M., (1972), *The Use of the Scanning Electron Microscope*, Pergamon, Oxford

Hirsch, P. B., (1960), in *X-ray diffraction by Polycrystalline Materials*, (eds H. S. Peiser, H. P. Rooksby and A. J. C. Wilson) Chapman and Hall, London, 278

Hirsch, P. B., and Humphreys, C. J., (1970), *SEMS*, 449

Joy, D. C., Booker, G. R., Fearon, E. O., and Bevis, M., (1971), *SEMS*, 497

Joy, D. C., and Newbury, D. E., (1971), *SEMS*, 113

Joy, D. C., and Jakubovics, J., (1968), *Phil. Mag.*, **17**, 61

Kanai, Y., and Uchibori, K., (1969), *SEMS*, 319

Lane, W. C., (1970), *SEMS*, 43

Lardner, E., and Boucher, N. A., (1972), private communication (Wickman Wimet Ltd, Hard Metal Division, Coventry, U.K.)

Newbury, D., and Yakowitz, H., (1973), *Proceedings of the 19th A.I.P. Conference on Magnetism*, Boston, Mass.

Nixon, W. C., (1969), *Contemp. Phys.*, **10**, 71

Oatley, C. W., (1969), *J. scient. Instrum.*, Ser. 2, **2**, 742

Oatley, C. W., (1972), *Scanning Electron Microscopy*, Cambridge University Press

Oatley, C. W., Nixon, W. C., and Pease, R. F. W., (1965), *Adv. Electronics Electron Phys.*, **21**, 181

Pawley, J. B., and Hayes, T. L., (1971), *SEMS*, 105

Pelloux, R. M., Erhardt, K., and Grant, N. J., (1970), *SEMS*, 283

Philibert, J., and Tixier, R., (1969), *Micron*, **1**, 174

Plows, G. S., and Nixon, W. C., (1968), *J. scient. Instrum.*, Ser. 2, **1**, 595

Scully, J. C., (1970), *SEMS*, 313

Scurfield, G., and Silva, S. R., (1969), *SEMS*, 187

Siegbahn, K., Nordling, C., Fahlman, A., Nordberg, R., Hamrin, K., Hedman, J., Johansson, G., Bergmark, T., Karlsson, S. E., Lindgren, I., and Lindberg, B., (1967), in *ESCA Atomic, Molecular and Solid State Structure Studied by means of Electron Spectroscopy*, Almquist and Wiksells Boktryckeri AB, Uppsala, chapter 6

Thornley, R. F. M., (1960), Ph.D. Thesis 'New Applications of the SEM', Cambridge University

Thornton, P. R., (1968), *Scanning Electron Microscopy*, Chapman and Hall, London

van Essen, C. G., Schulson, E. M., and Donaghay, R. H., (1970), *Nature*, **225**, 847

Waldron, G. W. J., Inckle, A. E., and Fox, P., (1970), *SEMS*, 297

Wells, O. C., (1974), *Scanning Electron Microscopy*, McGraw-Hill, New York

Yakowitz, H., (1964), in *The Electron Microprobe*, (eds T. D. McKinley, K. F. J. Heinrich and D. B. Wittry) Wiley, New York, 417

3

Microanalysis

The chemical analysis of an unknown or only partially identified material is a problem frequently encountered both in research and in industry. Some examples are the identification of fine precipitates in a metallic alloy or of inclusions in a synthetic fibre, the development of new alloys, the study of phase transformations in metallic or ceramic systems and the control of the specification of a material during its production. There are also numerous forensic applications, such as the analysis of a component involved in an industrial accident to determine whether it met the designer's specifications, or the identification of a fragment of paint from a painting that is suspected of being a forgery.

A very large number of chemical and physical methods of analysis have been developed to match these needs. Traditional 'wet' chemical analysis, chromatography, ion-exchange methods, spark mass spectrometry, neutron activation analysis and atomic absorption spectrography are some examples (a comparative survey is given by Henry and Blosser, 1970). However, the majority of these techniques give an average analysis over the whole sample and do not distinguish between, say, an element in solid solution in the bulk of a grain and the same element segregated to a grain boundary or a surface. Such information is essential to microstructural studies and in keeping with the theme of this book we shall only discuss those techniques capable of a fine-scale correlation between the chemical analysis and the microstructural features; that is, microanalytical techniques.

Some analytical information is available from the optical and transmission electron microscopes if the basic material is well characterised. A well-known example is the calculation of the carbon content of a plain carbon steel from the ratio of the areas of ferrite and cementite measured on a typical cross-section; also certain phases can be identified by their etching characteristics or by their electron diffraction patterns. The limitations of these methods are obvious. This chapter is concerned with the instruments and techniques that can provide qualitative and quantitative analysis of microstructural features that are perhaps less than 1 μm in diameter.

3.1 PRINCIPLES AND BASIC DESIGN OF MICROANALYTICAL INSTRUMENTS

Four techniques are currently available that can provide qualitative and quantitative analysis of a wide range of elements on a microscopic scale. Electron-microprobe analysis is by far the best developed and most widely used, and the bulk of this chapter will be devoted to it. It does, however, have certain disadvantages, some of which are overcome by newer techniques: Auger electron spectroscopy, secondary ion emission analysis and laser-microprobe analysis. These methods will be outlined and illustrated by some of their successful applications, but the design of the instruments will not be treated in any detail since their technology is still developing very rapidly. A fifth instrument, the atom-probe field–ion microscope, is capable of identifying the individual atoms that can be imaged in the field–ion microscope. From the nature of the specimens, however, this can never become an analytical technique in the sense used above and it is discussed, along with the field–ion microscope, in chapter 7 rather than in this chapter.

3.1.1 Electron-microprobe Analysis

The electron optics of the electron microprobe (often abbreviated to 'microprobe' or 'microanalyser') are generally similar to those of the scanning electron microscope, described in chapter 2. In section 2.5.3 the principle of the method was described, as an extra operating mode of the scanning electron microscope. Up to the present the best results in either scanning electron microscopy or in microanalysis have been achieved using separate instruments; the X-ray spectrometer on the scanning microscope and the scanning image facility on the microanalyser have been used only as auxiliaries. It is possible to design satisfactory combined instruments, and these are beginning to appear on the market, although they are not significantly cheaper than two separate instruments. There are a number of relatively minor differences between the design of a microprobe and a scanning electron microscope; for example a microprobe requires a higher beam current and higher maximum accelerating voltage to obtain sufficient X-ray intensity for accurate measurement and to excite the higher energy characteristic X-ray lines. There are, in addition, some more exacting requirements in a microanalyser: it is useful to be able to view the specimen with an optical microscope during the analysis, the X-ray geometry and the specimen position must be carefully designed to ensure good spectrometer performance, and several spectrometers should be fitted, each with as high a takeoff angle as possible. These requirements involve considerable redesign of the objective lens and the specimen chamber, and even when this is achieved there will be little room left in the specimen chamber for experimental stages (such as tensile testing jigs). Figure 3.1 shows a block diagram of a microprobe system in its modern form. The technique was invented and developed by Castaing in Paris. The design, performance and applications of microprobe analysis will be treated in later sections.

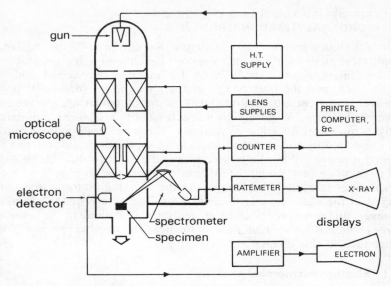

Figure 3.1. Block diagram of an electron-probe microanalyser

3.1.2 Auger Electron Spectroscopy

The principles of analysis utilising Auger electron emission were outlined in section 2.5.1. The microanalytical version of this method uses an instrument based on a scanning electron microscope (there are also instruments designed to use broad electron beams several square millimetres in area; these are of no interest for microanalysis). A device must be added to analyse the energies (velocities) of the emitted electrons, and suitable electronic circuits must be provided to extract the weak Auger signal from the general secondary electron signal. Since the Auger signal comes entirely from atoms within a few nanometres of the specimen surface an ultra-high vacuum environment is essential to maintain an atomically clean surface, and an ion gun is often installed as well, both to clean the surface and to erode the material to provide analysis in depth. Since an ultra-high vacuum must in any case be used, it is usual to provide a high-brightness field-emission gun (section 2.3.1) so that a good signal can be obtained with small probe diameters.

The vacuum requirement is not difficult to meet since the technology is now well established but the engineering is elaborate and the materials are costly, so that the final instrument is much more expensive than a scanning electron microscope. Electron energy analysers are straightforward in principle, simply being devices that force electrons with different initial energies into different trajectories. The most efficient analyser so far available, the electrostatic cylindrical mirror, is illustrated in figure 3.2. The energy of the electrons that follow the trajectory illustrated is determined by the potential on the outer cylinder, thus the analyser may be tuned to pass electrons of any energy.

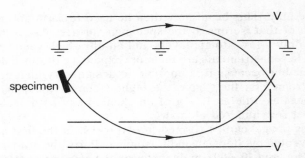

Figure 3.2. The electrostatic cylindrical mirror, an efficient electron energy analyser

As with electron-microprobe analysis, the electron beam may be focused on a point and a complete spectrum recorded, or the electronic circuits may be set to measure the intensity of electrons of a particular energy (corresponding to a particular element) and an 'Auger electron image' obtained by scanning the primary electron beam over the surface. Thus the technique may be regarded as one more mode of operation of the scanning electron microscope. Auger electron spectroscopy is, however, in an early stage of development and much of the work so far has been with systems capable of only a very crude correlation between the Auger signal and the microstructure. The possibilities are great, since the sampled volume in Auger spectroscopy is essentially the diameter of the electron beam and is only a few atoms deep, so that it may eventually be possible to obtain chemical analyses from regions less than 20 nm in diameter, and to achieve this resolution in a scanning Auger electron image. Technological problems, mainly in the design of the electron gun, currently limit the spatial resolution of a point analysis to about 100 nm, but this is already significantly better than any other form of microanalysis except the atom-probe field–ion microscope. The most important applications of Auger spectroscopy at present, however, are in the study of surfaces.

References to the theory of Auger spectroscopy were given in section 2.5.1. In sections 3.6 and 3.7.3 the analytic capabilities and the applications of the technique will be discussed in more detail.

3.1.3 Secondary Ion Emission Analysis

This is another relatively new technique for which production instruments are only just appearing. The principle is straightforward: a high energy (~ 12 keV) 'primary' ion beam, usually of A^+, O^- or O_2^- ions, is directed at the specimen surface. 'Secondary' ions are sputtered from the surface, in effect removing a very small sample of the surface for analysis. The secondary ions are accelerated by an electric field and passed into a mass spectrometer, which sorts the ions according to their mass/charge ratio. The beam leaving the spectrometer is thus tuned to a particular e/m ratio and with a good spectrometer will consist of ions of one isotope only. This beam may then be used to form an image of the distribution of that isotope at the specimen surface. Again the depth of analysis is only a few atomic layers and an ultra-high vacuum system is essential. These instruments are the most expensive of those described in this chapter, but because it is possible to design very-high-performance mass spectrometers, they have the highest sensitivities; in favourable circumstances as little as 10^{-19} g of an element can be detected (corresponding to a few hundred atoms).

The 'mass image' can be formed in two ways. In the first and original method, invented by Slodzian and Castaing in Paris, the primary ion beam illuminates an area 20–400 μm diameter on the specimen surface, and the sputtered ions are focused with ordinary magnetic lenses before and after the mass spectrometer, as shown in figure 3.3. Using a suitable image

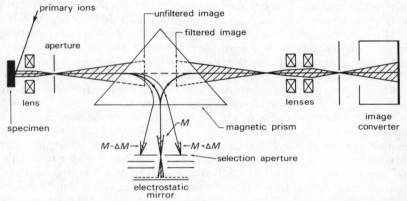

Figure 3.3. Block diagram of a secondary ion emission microscope. (Courtesy of G. Slodzian)

converter the isotope image can be displayed and photographed. Alternatively a method analogous to that of the scanning electron microscope can be used, in which the primary ion beam is focused to a fine probe (a few microns diameter at present); this is scanned in a raster over the specimen surface, and a signal from a detector placed at the output of the mass spectrometer modulates the brightness of a cathode-ray tube. The choice between the two methods depends mainly on whether the development work is put into ion optics or into scanning and detection electronics. Although it is easier to obtain high magnifications by the scanning method, it is rather difficult to produce a really fine high-intensity ion beam and the resolution of the original method is rather better at about one micron. However, there is no 'resolution race' between the two methods: the main research effort at present is, very appropriately, aimed at understanding the mechanisms of secondary ion production in order to improve the interpretation of the images and the accuracy of the analysis, rather than at much improvement of the technology.

In either design of 'ion microscope' an area a few microns in diameter may be selected and analysed completely. The surface is of course eroded during the analysis, at a rate of 1–100 nm s^{-1}, so the method is destructive; however, the change in composition with depth in the specimen may easily be studied.

A general review of ion microscopes is given in Socha (1971), and descriptions of the particular instruments in Slodzian (1964) Rouberol *et al.* (1968) and Liebl (1967). A recent conference report on ion-beam surface layer analysis has been edited by Mayer and Ziegler (1974).

3.1.4 Laser-microprobe Analysis

The spectral analysis of the light emitted from a specimen excited by a spark has long been used as a routine method of analysis. The energy of excitation is relatively low, and only the transitions between outer electron shells in the atom are possible. These cause the emission of radiation in the visible or near-visible band. The invention of the laser provided a means of microanalysis using this technique, since lasers can provide high intensities of incident radiation that can be focused to a spot about 2 μm diameter. When a laser pulse is focused onto a solid specimen a small volume of material is directly evaporated and simultaneously excited; the emitted radiation can then be analysed in a conventional optical spectrograph, recording the intensities of the spectral lines on a photographic plate. Sharper and more intense lines are obtained by providing auxiliary excitation of the vapour by means of a spark passing just above the surface of the specimen.

An outline of the arrangement of the laser microprobe is shown in figure 3.4. A ruby laser is normally used, the output of which is typically a

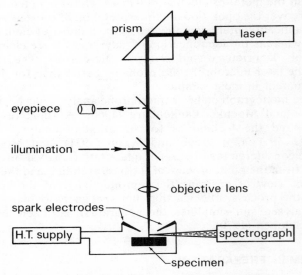

Figure 3.4. Block diagram of a laser-microprobe analyser

pulse of duration 30 ns with peak power about 4 MW. Reproducibility of the laser pulse is essential for quantitative analysis. Otherwise, the apparatus has the basic facilities of the optical microscope with the addition of electrodes and power supplies for the auxiliary spark, and, of course, safety interlocks to prevent the laser pulse arriving at the eyepiece.

Only spot analyses are available with the microprobe, and it is not possible to obtain images of the distribution of one component. Although the spot can be made reasonably small, analyses cannot be taken from closely adjacent regions if the specimen is analysed in air, since the residue and oxidation caused by the pulse generally spread several millimetres across the specimen surface. The contaminated area can be reduced by physical shielding or by providing an inert atmosphere around the specimen, but the lack of complexity of a vacuum system is a considerable advantage to this method of analysis.

Much of the experience gained with spark spectrographic analysis can be applied to the laser-microprobe method, and analyses accurate to about ± 10 per cent can be performed. Only 'metallic' elements can be analysed, because only metallic atoms can lose electrons (by impact ionisation) to form the ion–electron plasma that comprises a spark discharge. However, the material itself need not be metallic, and ceramics, polymers and organic materials can be successfully analysed for their metallic-element content. Another advantage of the technique is that, unlike methods based upon beams of electrons or ions, the electrical conductivity of the specimen is irrelevant and insulators can be examined as easily as metals.

While the laser microprobe is not as accurate or as versatile as the electron microprobe, its performance is adequate for many purposes especially in the metal-producing industries, and it is much cheaper and faster. Moreover the operation and interpretation of the analysis does not become significantly more complicated as the number of elements present increases, whereas the identification of, say, thirty trace elements would be extremely laborious on an electron microprobe. The detectability limits for the laser microprobe are probably better than for the electron-beam instrument in most samples.

A detailed monograph on laser-microprobe analysis has been published by Moenke and Moenke-Blankenburg (1966). A short review of the instrument and the techniques for quantitative analysis is given in Rasberry et al. (1967); Webb and Webb (1971), and Moenke et al. (1970) have considered the process of quantitative analysis in more detail, the former from the point of view of the interpretation and the latter from the point of view of ensuring the reproducibility of the apparatus. A variant of the process, utilising thin-film specimens (which reduces the volume analysed and simplifies the apparatus) has been described by Mela and Sulonen (1970).

3.2 SPECIMEN PREPARATION

For some qualitative or semi-quantitative applications the preparation of the surface is not critical and can follow the routines of metallographic or petrographic preparation. In some cases, the nature of the specimen may already be dictated, for example if one is examining a fractured surface in a scanning electron microscope and wishes to analyse an inclusion suspected of having initiated the fracture. For good quantitative work, however, the surface preparation is important, and in the instruments that analyse a layer only 2 nm deep it is clearly essential to produce a clean, representative surface. Two constraints on specimen preparation can be distinguished

(a)   The surface smoothness must permit the instrument to operate at full efficiency and accuracy; that is, the surface must have good 'optical' properties.

(b)   The surface must be truly representative of the bulk material if microstructural features common to the whole material (rather than just to the surface) are being examined.

Good geometric or optical qualities are required in electron-microprobe analysis, for two reasons. First, the most accurate X-ray spectrometers are of the focusing type (section 3.3.2), which must be accurately aligned with the X-ray source (the point at which the electron beam strikes the specimen). As illustrated in figure 3.5, if there is too much variation in

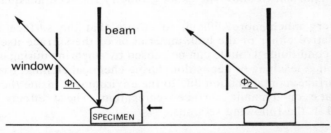

Figure 3.5. The effect of specimen roughness on the apparent X-ray source position; the source moves vertically as the specimen is translated

surface height the source will shift whenever the specimen is translated, and the apparent efficiency of the spectrometer will deteriorate. Secondly, spurious X-ray signals can be produced on a very rough surface as shown in figure 3.6. A surface projection may intercept and absorb some of the X-ray signal from the point under analysis, as at A in the figure, or it may fluoresce under the effect of this signal (or of backscattered electrons) and add to the measured X-ray signal a component that has nothing to do with the area under the electron beam, as at B.

As a general rule, surface roughness and lack of flatness will only affect the precision of analysis if the elevation differences on the specimen surface are larger than the diameter of the electron beam ($\sim 1 \mu m$). As might be expected the problem is more acute if the takeoff angle is low (this is the angle between the specimen surface and the direction in which the X-rays enter the spectrometer). Even so, careful metallographic polishing ending with abrasive particles $\sim 0.5 \mu m$ in diameter should suffice.

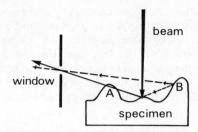

Figure 3.6. The effect of specimen roughness on the spurious absorption or generation of X-rays. Spurious absorption is caused at A and spurious generation, stimulated by backscattered electrons or fluorescent X-rays, occurs at B

The other analytical methods do not at present require such a flat surface, partly for the negative reason that quantitative measurements by these methods are not yet well developed and the effect of surface roughness is not so well understood. Conventional metallographic polishing is certainly adequate for the laser microprobe. The flatness requirements for Auger spectroscopy and ion emission microscopy are not stringent at present and fractured surfaces are commonly used, but it will become important to know the exact geometry for accurate quantitative analysis.

It is very much more difficult to ensure that the surface is really representative either of the bulk material or of the surface itself under ordinary conditions. Trouble can be caused by smearing during mechanical polishing; leaching or deposition during chemical etching; oxide films on the surface; contamination due to the environment inside the microscope; static or dynamic surface segregation. These effects will be discussed in the following sections.

3.2.1 Effects of Polishing and Etching

The polishing problem occurs when a specimen contains more than one phase. The mechanical forces during polishing can smear a layer of one phase over another (especially if a very soft phase is present) giving a thin surface layer quite unrepresentative of the bulk material. Even with the more penetrative methods of electron-microprobe and laser-microprobe analysis, layers one or two atoms deep can affect the analysis in unfavourable cases, and it is essential that the smeared layer should be removed by chemical or electrolytic etching. This must, however, be as light as possible since prolonged etching can give two harmful effects, namely deposition and leaching. Deposition is caused by local galvanic action at the specimen surface. It can be minimised by keeping the etching time short, but it is much better to prevent it completely by using etchants that only contain ions that are less noble than the elements in the specimen itself; for example a specimen containing iron should not be etched with a solution containing copper. Leaching is the preferential dissolution of one or more components in the specimen and again results in a surface untypical of the bulk material or even of the original surface. The rate of leaching will be controlled by the rate of supply of the dissolving atoms to the surface, that is, by diffusion in the solid. It is therefore best to choose an etchant whose action is rapid (as well as roughly equal on each component). As in the case of smearing, deposited or leached layers only 0.1–10 nm deep can sometimes affect the accuracy of the analysis.

3.2.2 Surface Oxide Films

All oxide or other surface films must be rigorously excluded in Auger or secondary ion methods of analysis, as discussed below. As long as they are not too thick, oxide films will not affect laser-microprobe analysis since this method is not sensitive to oxygen. In electron-microprobe analysis any surface film can affect the analysis in the same way as a smeared, deposited or leached layer, although a few atomic layers of oxygen can in this case be tolerated since it is a light element. Accidental oxidation (perhaps by anodic oxidation during electropolishing or elec-troetching) where the thickness and composition of the film is unknown, must be removed by etching. An example is given in section 3.7.1 of an aluminium alloy in which the magnesium content of an 'oxide' film was found to be 1000 times greater in the film than in the bulk material, and the formation of such mixed oxides can obviously distort the whole analysis.

A rather different situation occurs when a controlled and well-defined oxide film (up to 150 nm thick) is created on the surface to increase the contrast in optical microscopy. A film of this thickness does not accumulate enough charge to affect the electron beam. Although it is thick enough to affect the analysis, quantitative allowances may be made for a well-defined film, either by calculating its effect upon the generation and absorption of X-rays or by coating the standard used for comparison of the X-ray intensities with an identical film.

3.2.3 Contamination within the Instrument

The laser-microprobe method is the least sensitive to environmental contamination; indeed, analyses are usually performed in the atmosphere, although more spot analyses can be taken from one specimen if the surface is protected during the evaporation process by the use of a vacuum or inert gas. In the case of electron-microprobe analysis, carbon contamination used to be a serious problem; the oils used in the diffusion and rotary pumps of a conventional vacuum system can become cracked in the electron beam and deposit a layer of carbon on the specimen surface, affecting the accuracy of analysis and the visibility of features. However, this problem has been virtually eliminated by better vacuum technology: better oils and greases, higher vacua and better anti-contamination stages (usually liquid nitrogen traps). If a modern micro-probe (or scanning or transmission electron microscope) shows carbon contamination problems it is probably faulty, badly maintained or incorrectly operated.

The problem of contamination is very much more severe for the surface-sensitive techniques of ion emission analysis and Auger spectroscopy, since the surfaces must be perfectly clean on an atomic scale. The surface must be prepared within the ultra-high vacuum of the microscope itself since oxygen monolayers form very rapidly; even at 10^{-6} torr a monolayer will form in a period of time much less than that required to perform an Auger or secondary ion analysis, and vacua in the 10^{-8}–10^{-11} torr range must be used (the requirements for Auger analysis are in fact at the low-pressure end of this range while those for secondary ion analysis are a little less severe since the surface is continuously cleaned by the primary ion beam and the time of analysis is considerably shorter). The most satisfactory method for preparing the surface is by fracturing the specimen within the microscope, where practicable. Failing this, the surface can be cleaned by sputtering the surface layer off with the primary ion beam or an auxiliary ion gun or by heating the specimen to desorb the adsorbed gas layers. However, these methods carry the danger of preferential sputtering of certain components or (if the specimen is heated) of preferential evaporation and of surface segregation of impurities from the bulk. The problems outlined in this paragraph must be taken very seriously by practitioners of surface analysis, and advanced treatments may be found in books such as that on clean surfaces edited by Goldfinger (1970) and that on ultra-high-vacuum techniques by Redhead *et al.* (1968).

3.2.4 Surface Segregation and its Detection

In some instances, the surface may be untypical of the bulk material for good physical reasons rather than through any defect in the specimen preparation or environment, and it may well be important to study these differences. Grain boundary segregation may appear as surface segregation if the specimen is fractured along a grain boundary; inverse segregation in a metallic alloy casting may cause a relatively thick surface layer; archaeological specimens may have effectively suffered natural leaching for centuries; finally, surfaces are simply different from the interior of a solid and for example surface segregation may exist for energetic reasons similar to those causing grain boundary segregation (easier nucleation of a new phase and higher diffusion coefficients relative to the bulk). It is often possible to detect untypical surface layers by analysing in depth. In the electron microprobe this is performed by altering the accelerating voltage of the primary ion beam; increasing the voltage increases the penetration of the electron beam and the mean depth of X-ray production increases. A change in the apparent concentration of any element as the accelerating voltage changes is therefore a good indication of an untypical surface layer. In the ion-emission microscope the specimen is continuously eroded by the primary ion beam and a change in apparent concentration with time gives a similar indication. A comparable test may be made with

Auger spectroscopy (or, on a larger scale with laser-microprobe analysis) by eroding the surface with an auxiliary ion gun if preferential sputtering is not a problem. An example of this type of application is given in section 3.7.3, in a discussion of experiments in which a specimen is fractured along a grain boundary and a depth analysis performed to determine the amount and extent of grain boundary segregation.

A much more serious problem is presented by dynamic surface segregation. An example was mentioned in the last section, the segregation of impurities from the bulk to the surface on heating the specimen, but the possibility exists whenever an element is so mobile in the lattice or on the surface that its concentration can change appreciably at the surface during the time required for setting up and performing the experiment. Even fracturing the specimen will not then ensure a surface typical of the bulk. One would expect lattice mobilities high enough to cause complications in such cases as interstitial solutions in body-centred cubic metals (for example, hydrogen or carbon in iron) but there are some indications that even elements with low diffusion coefficients in the bulk are highly mobile on the surface. Similar problems can be caused if the vacuum around the specimen is insufficiently high; this, however, can be cured by better vacuum technique whereas the difficulties of characterising and overcoming dynamic surface segregation are by no means solved, partly because they are at present poorly understood.

3.2.5 Insulating Specimens

The electrical conductivity of the specimen is quite irrelevant in laser-microprobe analysis, but insulating specimens cannot be used for the other methods. As discussed in section 2.2 for scanning electron microscopy, charge from the primary ion or electron beam accumulates on the surface and eventually deflects the beam itself. This problem can be overcome by coating the specimens with a conducting layer. For the electron microprobe the coating is usually a continuous film of a light element (carbon or aluminium), which minimises the absorption of the electron beam and of the characteristic X-rays. A similar film is applied to the reference standard, or alternatively a plain standard is used and the effects of the film on the specimen allowed for by calculation. The coatings used for Auger or ion microscopy cannot, of course, be continuous but if they are laid down in the form of a grid they are still effective in removing the charge accumulation while the interstices of the grid present a clean surface for analysis. Alternatively, the rate of arrival of electrons (in an electron beam or negative ion beam) can be balanced by the electron emission from the specimen through correct choice of the primary beam energy. In practice this means operating an electron-probe instrument at a lower operating voltage than usual and accepting the fact that only X-rays of longer wavelength can then be excited.

Thermal and radiation damage can also occur when investigating insulators; minerals are usually strong enough to withstand damage but serious problems can occur with inorganic salts, polymers or biological materials. The best conditions must be chosen for each individual specimen since reduction of the damage by lowering the beam energy and intensity also results in reduced sensitivity and accuracy of the analysis and an increase in the exposure time needed for a satisfactory determination. It is sometimes useful to cool the specimen, if necessary with a liquid helium stage. Secondary ion emission microscopy will probably become more important for these difficult materials because of its much higher basic sensitivity, especially for light elements (which comprise a large proportion of materials such as polymers and biological tissues).

A detailed discussion of the methods of preparation of metallic specimens for electron-probe microanalysis has been given by Hallerman and Picklesimer (1969).

3.3 DESIGN OF THE ELECTRON MICROPROBE

This section is restricted to the electron microprobe, since it alone among the instruments described in this chapter has been thoroughly developed and widely used. Much of the design is based upon principles already discussed in chapter 2, and the following sections will discuss mainly the differences in design of the SEM and the microprobe; however, the design and performance of X-ray spectrometers used as SEM accessories will also be treated.

3.3.1 Electron and Light Optics

The electron optics are generally similar to those of the SEM. In the past, microanalysers were designed with relatively crude electron optics (for example, they may have been provided with only one electron lens) since the spatial resolution in X-ray microanalysis is limited to about 1 μm by the spread of the beam below the surface (sections 2.5.1 and 2.5.3). Instruments were therefore designed to give an electron beam no finer than 0.5 μm (compared with 10 nm in the SEM) and no great precautions were taken to exclude stray magnetic fields or stage vibrations. The secondary electron images were consequently poor. There is now a tendency to build instruments that are equally good in either application in which case the electron optics are naturally very similar to those in the SEM.

The microprobe has traditionally included an optical microscope to view the area under analysis. This is partly because the development of microanalysers and scanning electron microscopes was, surprisingly, separate; only now are the two beginning to merge. With good secondary electron images there is no real need for an optical microscope but it does provide some extra contrast methods such as interference contrast, polarised light and cathodoluminescence (the eye is a cheap detector of visible light), which help in correlating the compositional analysis with the microstructure. The optical microscope is indicated schematically in figure 3.1; there are several ways of combining an electron lens with an optical lens (see for example Tousimis and Marton, 1969) although separate electron and optical axes are used in some instruments, together with an accurate stage-traversing device to align a given point on either axis.

The conventional magnetic lens with an iron core is not necessarily the best choice for a microanalyser, because the iron core makes the lens so bulky. Space is very restricted near the objective lens of a microanalyser, and some modern instruments make use of the 'minilens' invented by Le Poole (1964). This lens dispenses with the iron core and uses a plain coil with relatively narrow radius. The coil must be accurately wound to avoid excessive astigmatism, but the final design is very compact and permits, for example, an increase in the X-ray takeoff angle and better optical microscope facilities. It is also used as an extra lens in some transmission electron microscopes to obtain a very fine spot for accurate selected-area diffraction from small particles, and as a standard component of the combined electron microscope and microanalyser, EMMA, described in section 3.4. Duncumb (1969) discusses the applications of the minilens in more detail.

3.3.2 X-ray Spectrometers

The two principal types of spectrometer, wavelength-dispersive and energy-dispersive, were outlined in section 2.5.3. We shall now examine their design and characteristics in more detail.

Wavelength-dispersive spectrometers work by diffracting the X-ray beam from a crystal of known interplanar spacing and measuring the Bragg angle and intensity of the possible diffracted beams. To maximise the peak reading for any diffracted beam the spectrometer must

(1)   Collect the X-rays at a high takeoff angle to minimise absorption in the specimen.

(2)   Collect the X-rays over as high a solid angle as possible.

(3)   Minimise losses in the diffracting crystal and in the whole path between the specimen and counter.

(4)   Use an efficient counter to measure the X-ray intensity.

(5)   Have ample gain and accuracy available in the measuring electronics.

It is also important to maximise the peak-to-background, or signal-to-noise ratio. To achieve this it is necessary to

(1)   Ensure that no spurious X-rays enter the spectrometer (for example from backscattered electrons striking parts of the specimen chamber).

(2)   Use a crystal with the sharpest possible peak of X-ray intensity *versus* angle of incidence when the crystal is rotated through the Bragg angle (that is, the narrowest possible rocking curve).

(3)   Minimise the noise in the counter and its associated electronics.

X-ray counters and the measuring electronics are described in section 3.3.3. The high takeoff angle requirement is a matter of geometry and electron optics, and helps to explain why the design of an instrument equally good as an SEM and as a microanalyser is difficult, since the ideal position for the spectrometer is in the middle of the objective lens (already occupied by the optical microscope). Takeoff angles are in the range 35° to 75° in most commercial instruments. Careful design is needed to ensure that no spurious signals enter the spectrometer and absorption is minimised by evacuating the whole spectrometer and using a thin window of an atomically light material (beryllium or a plastic such as Mylar) between the vacuum and the X-ray counter.

The solid angle of X-ray collection is usually rather low in wavelength-dispersive spectrometers since they are bulky and cannot physically be placed very close to the specimen. It is therefore essential to ensure that the highest possible proportion of the X-rays actually entering the spectrometer also enters the X-ray counter after diffraction by the crystal. This is achieved by using the focusing action resulting from an elementary geometric property of a circle, that angles based upon the same arc are equal. If the X-ray source, diffracting crystal and counter all lie on the circumference of a circle (the 'Rowland circle') this condition is achieved. The crystal will diffract over its whole length if it is elastically bent to a radius $2R$, where R is the radius of the Rowland circle, and its surface ground to a radius R. These conditions are illustrated in figure 3.7. The crystal and counter move around the circle to analyse different wavelengths and if the source is stationary and the takeoff angle fixed, the centre of the circle must also move. This 'fully focusing' principle is often called after its inventor, Johansson. It is possible slightly to relax these conditions and still have an excellent spectrometer; in the linear spec-

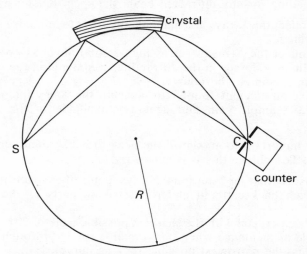

Figure 3.7. The Johansson principle for focusing spectrometers. The X-ray source S, the crystal and the counter entry slit C lie on the circumference of the Rowland circle, radius R. The crystal is bent to a radius $2R$ then its surface ground to a radius R so that it lies exactly on the circumference of the Rowland circle

trometer design used in many commercial microprobes, the X-ray counter slit does not remain exactly on the Rowland circle but the performance hardly suffers and the spectrometer is particularly compact and convenient to use. If the conditions are relaxed further—for example, in some spectrometers the crystal and detector rotate about a fixed axis—a semi-focusing spectrometer results. These used to be quite popular for work with rough surfaces, since a fully focusing spectrometer demands an accurately fixed source position in order to be fully focusing (hence the flatness requirement discussed in section 3.2; see figure 3.5). However the efficiency of semi-focusing spectrometers is poor and they have been superseded by the energy-dispersive type discussed below.

A range of crystals is needed to cover the range of wavelengths that has to be analysed (from about 20 pm to 15 nm). Low absorption and high perfection are needed and a large number of crystals is available, from lithium fluoride ($d = 89$ pm) to complex organic salts such as lead lignocerate ($d = 6.3$ nm) and lead melissate ($d = 8.0$ nm). Pseudo-crystals have also been used, consisting of alternating layers of different metals or even of soap films (stearates). The main problem is in finding crystals that are useful at the longer, easily absorbed wavelengths. Gratings cut on glass (utilising the total reflection of X-rays at low angles) are being used with considerable success, although since a different spectrometer design is needed gratings cannot be interchanged with crystals. A good, brief survey of crystals and gratings is given by Beaman and Isasi (1972).

Energy-dispersive spectrometers work on the principle that the energy of a photon, E, is related to the frequency of the electromagnetic wave, ν, by the relationship

$$E = h\nu \qquad (3.1)$$

where h is Planck's constant. Moseley's equation (2.21) can therefore just as well be formulated in terms of energies

$$\sqrt{(E/h)} = Z - C \qquad (3.2)$$

and measurement of the photon energy identifies the element concerned. To some extent the proportional counter (section 3.3.3), normally used as the counter in a wavelength-dispersive spectrometer, will act as an energy-dispersive detector on its own since it does produce pulses proportional in energy to the incident photon energy. However, its resolution (discrimination between photons of slightly differing energies) is poor and high-resolution detectors are made from silicon that has been 'drifted' (doped) with lithium. In section 2.5.4 it was pointed out that the primary electrons will induce electron–hole pairs in the specimen by impact ionisation. A photon beam will have a similar effect, and the number of ion pairs, n is given by

$$n = E/E' \qquad (3.3)$$

where E' is the energy required to create an ion pair. Measurement of n thus gives the photon energy E (E' is 3.8 eV for Si(Li) at 77 K). This is achieved by applying a bias voltage across the detector, so that the electrons and holes move to opposite sides of the slice, and measuring the total charge flow Q from the relationship

$$Q = \int_0^{t'} i \, dt \qquad (3.4)$$

that is, the time integral of the current over the duration t' of the pulse. Division of Q by $2e$ where e is the electronic charge gives the number of ion pairs, and equation 3.3 can then be used to calculate E. A high input impedance, high gain and low noise are essential in the amplifier used to measure the charge; the eventual output is a voltage pulse where the voltage is proportional to the photon energy.

Photons of all the energies corresponding to the emitted X-ray spectrum are falling on the detector almost simultaneously, and the measuring process must be very rapid to avoid confusion. From necessity therefore, the energy-dispersive spectrometer offers the advantage that all wavelengths in the range of the device can with suitable electronics be analysed and displayed virtually simultaneously. The output voltage pulses are passed to a multichannel analyser having perhaps 1000 channels, each allocated to a given voltage range. When a voltage pulse arrives it is allocated to the channel appropriate to its value and the count stored in that channel is increased by one. The multichannel analyser thus stores the whole spectrum as it arrives and an accurate spectrum can be obtained in seconds or minutes, compared with hours for the wavelength-dispersive spectrometer (which can only examine one wavelength at a time while photons at other wavelengths are wasted). The spectrum shown in figure 3.15c was obtained in this way. The collection of X-rays is very efficient with the silicon detector, which is 5–25 mm diameter and can be placed very close to the specimen to collect X-rays over a large solid angle. Its major inconvenience is that the detector must be kept permanently at a low temperature since it deteriorates at room temperature. A liquid nitrogen cryostat must be provided (outside the microscope column) and kept filled at all times.

Before comparing the performance of the two types of spectrometer it is best briefly to consider the factors that influence the performance of any spectrometer. The first is the peak reading, P, obtained when counts from a given peak are collected over a given period. This is controlled by the efficiency of collection of X-rays: the solid angle of collection, the amount of absorption in the windows and elsewhere, and the basic efficiency of the collector. A high P value indicates a high sensitivity and hence easier operation at low beam currents. It also controls the speed at which an analysis may be performed—decidedly not a trivial factor in practice. Secondly, the peak-to-background ratio, P/B; this deteriorates if spurious X-rays are detected or if there is any inherent noise in the detector or its associated electronics. High P/B values give accurate analysis through accurate measurements of peak or integrated intensities and also provide a lower limit of detectability for a given element (this is controlled by the ratio P^2/B for the X-ray line characteristic of that element, as shown by Ziebold, 1967). The third important factor is the resolution, usually defined as the width of a given X-ray line measured at half its peak height. This may be quoted in wavelength, frequency or energy units; the last is common, normally in electron volts. Resolution is important both for distinguishing elements with closely separated characteristic X-ray lines, and for making measurements on a line free from any interferences from other lines. Interferences can occur even when the peaks themselves may be distinguished, as illustrated in figure 3.8. Finally, we must

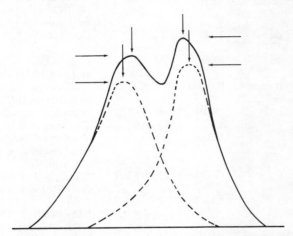

Figure 3.8. An illustration of interferences occurring even when spectral lines are distinguishable. The dashed lines are the true shapes of two adjacent peaks and the solid line is their sum. The peak positions and heights are each slightly modified by the presence of the other peak

consider the performance of the spectrometer when perfectly flat specimens are not available. Bearing these points in mind, table 3.1 may be used to compare various aspects of the performance of the two spectrometers. Where possible, numerical values are quoted but it must be remembered that these are only for comparison and for inculcating a feeling for the values, since these will vary widely between different instruments and operating conditions. A detailed comparison of spectrometers can be found in Beaman and Isasi (1972).

In summary, the two types of spectrometer are complementary rather than competitive. The energy-dispersive spectrometer can do three things that the wavelength-dispersive type cannot do at all: obtain a spectrum very rapidly, obtain useful results at low beam currents (essential for fragile specimens) and obtain analyses from rough surfaces. On the other hand, the wavelength-dispersive spectrometer has the higher accuracy and resolution, lower detectability limits and better performance with light elements. The choice of spectrometer thus depends on the application: a wavelength-dispersive type where the emphasis is on the best possible analysis, an energy-dispersive type where analytical information is a valuable adjunct to other studies or where semi-quantitative results

Table 3.1. A comparison of wavelength-dispersive (WDS) and energy-dispersive (EDS) spectrometers
(The numerical values are for illustration only; they will vary with the manufacturer, spectrometer geometry, wavelength and specimens used)

Factor	WDS	EDS	Reason for difference
Time for complete spectrum (min.)	25–100	0.5–5	(1) Collection efficiency (2) EDS measures whole spectrum simultaneously
Count rate on one peak (c.p.s./nA)	1000	10000	Collection efficiency
Peak/background ratio	1000	50	EDS collects spurious X-rays and has high inherent noise
Maximum count rate (c.p.s.)	100000	30000	EDS counts all channels simultaneously, and saturates
Resolution, 0.1–10 keV (eV)	5–10	150–200 80–100	Currently possible Theoretical limit. WDS inferior to EDS above 25 keV, (suitable crystals not available)
Detection limits (weight ppm)	50–1000	2000–5000	WDS used at higher beam current, fewer overlapping peaks EDS better if current restricted, e.g. to avoid beam damage
Accuracy of analysis (per cent)	±1–2	±6	Experimentally determined
Light element analysis (min. atomic no.)	4	10	EDS uses Be window, WDS uses collodion
Rough surface work	Bad	Good	EDS insensitive to source position

are all that is needed (as is often the case). It follows that the obvious choice for an attachment to the scanning electron microscope is the energy-dispersive type, but that microprobe analysers are normally fitted with several wavelength-dispersive spectrometers with, if possible, an energy-dispersive spectrometer as well to give a really fast preliminary survey of an unknown specimen.

3.3.3 X-ray Counters and Counting Electronics

Energy-dispersive detectors themselves form efficient counters, but they are not often used as the detector in a wavelength-dispersive spectrometer because of their expense, external weight and bulk (caused by the cryostat) and inefficiency at low energies (light elements) caused by the absorption in the beryllium windows. Instead one requires a detector that is lightweight, accurate, highly efficient over a wide range of wavelengths, and capable of counting at a very high rate. It is also useful if the counter has some energy-dispersive properties so that higher-order reflections (with correspondingly shorter wavelengths) at a given Bragg angle can be rejected. However, the energy resolution does not need to be very high for this purpose, say 1000 eV.

The proportional counter is almost always used on wavelength-dispersive spectrometers. It is simply a development of the well-known Geiger counter, in which the ion–electron pairs created in the working gas (argon or xenon) are accelerated towards the wall of the counter chamber by a suitable voltage, and a voltage pulse is obtained. The principle is similar to that of the solid-state detector used in energy-dispersive spectrometers. In the Geiger counter, however, the accelerating potential is so high that as the ions move towards the chamber walls they cause an 'avalanche' of further ionisation. This results in a very strong pulse, which can be detected with simple equipment, but which has to be quenched before a further pulse can be received (either by reducing the voltage or by the presence of certain additives in the gas). The proportional counter differs from the Geiger counter mainly in that its accelerating voltage is much lower. Less avalanche ionisation is obtained and a weaker pulse results, but the pulse is now proportional to the initial photon energy and is quenched much faster so that the 'dead time' of a proportional counter is about 200 ns compared with 50 μs or more for a Geiger counter. The efficiency and resolution of a typical proportional counter are shown in figure 3.9. A usable sensitivity remains even at wavelengths of 10–20 nm; at such wavelengths an extremely thin plastic window must be used, or the window eliminated entirely (if the vacuum system of the microprobe can cope with the resulting input of gas). A continuous flow of gas can be used for very high count rates, effectively giving a very rapid quench of the pulse by physically removing the ionised gas.

Scintillation counters also produce a pulse proportional to the initial photon energy. They are similar in principle to the detector of a scanning electron microscope, consisting of a scintillation crystal (sensitive to photons in the X-ray band, rather than electrons) plus a photomultiplier. They are very efficient, and can count at even higher rates than proportional counters, but their performance drops off badly at wavelengths longer than about 200 pm so they are not normally used in wavelength-dispersive spectrometers. Their main use is in gamma-ray spectroscopy and in X-ray diffraction; they are ideal for monitoring diffracted beams when adjusting a crystal for X-ray topography (see chapter 6).

The electronic systems used to count the pulses from the detector are very important. The normal process is to pass the (energy-proportional) pulses through a pulse height analyser to reject signals from X-rays of different wavelengths, then into a ratemeter, which indicates the rate of arrival of pulses, or into a counter, which is used for accurate work and which counts the number of pulses received in a given time. There may also be various refinements such as a pulse-shaper, which alters the initial shape of the pulse to one that is better accepted by subsequent devices. The whole range of count rates is found in electron-beam microanalysis, from a few counts per second up to hundreds of thousands.

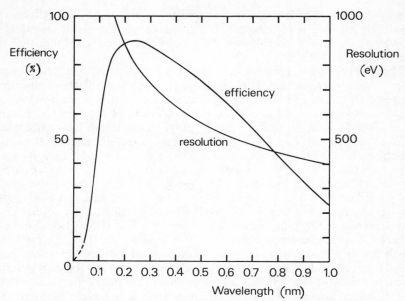

Figure 3.9. Characteristics of a proportional counter: counting efficiency and resolution as functions of wavelength. The curves are appropriate for a xenon-filled counter but depend greatly on the filling gas, the absorption in the window and the design and operating voltage of the counter. The long-wavelength performance can be improved by using argon instead of xenon and by using a very thin window or a windowless counter

The random nature of the emission of electrons from the gun, and hence of the emission of X-rays from the specimen, causes statistical noise in X-ray microanalysis as in scanning electron microscopy (section 2.4.2). This has several effects. First, the images formed by modulating the CRT from the signal due to characteristic emission from a particular element are very noisy (see for example figure 3.13) as discussed briefly in section 2.5.3. Second, there will be bursts of pulses at much higher rates than the average and the counter will 'drop' counts at lower rates than it would if the pulses were evenly spaced. Third, there will be errors, due to the statistical noise, that will be more acute at lower total counts. Equation 2.6 may again be used to express a 99.7 per cent confidence that two signals are different and in this case the distribution follows Poisson statistics quite accurately, so that the standard deviations may be taken as the square root of the total number of counts in each signal. If there are n counts in the first signal and the second signal has a mean value $(1 + C)$ times that of the first signal, equation 2.6 may be rearranged to give the number of counts required to distinguish two signals whose 'contrast' is C

$$(n)^{\frac{1}{2}} = \frac{3[1 + (1 + C)^{\frac{1}{2}}]}{C} \tag{3.5}$$

therefore

$$n \approx \frac{36}{C^2} \tag{3.6}$$

if C is small. The choice of a 99.7 per cent confidence level is of course arbitrary (90 per cent is often used; this would change the numbers 3 and 36 in the above equations to 1.65 and 10.89 respectively), but using this very high confidence level the number of counts needed to distinguish between two signals is plotted in figure 3.10. Clearly, C may also be taken as a measure of the fractional precision of a reading. From figure 3.10, it can be seen that to attain a 99.7 per cent confidence the errors in a count are no larger than 0.1 per cent ($C = 0.001$) some thirty-six million counts must be accumulated.

It is common to use computers to process microanalytical results. Initially their use was restricted to the application of the corrections for quantitative analysis (section 3.5) but microanalysers are now often fitted with a small 'dedicated' computer. This can process and store the results and can also perform some of the analysis, such as repetitive movement between standard alloys and various parts of the specimen (previously

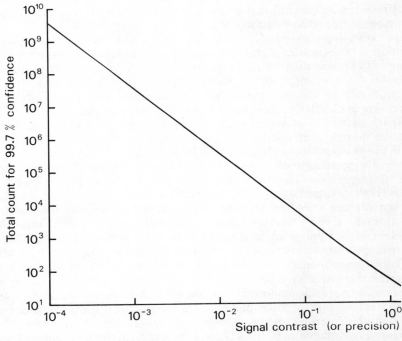

Figure 3.10. Counting statistics; the graph shows the number of counts n required to distinguish (with 99.7 per cent confidence) between two signals whose fractional difference (contrast) is C. C may also be interpreted as the fractional precision of a reading comprising n counts

specified by the operator). This can be very helpful in the repetitive analysis of simple, well-defined samples; obvious examples are the measurement of diffusion profiles in otherwise homogeneous material, or the analysis of hundreds of similar particles in an air-pollution study, but most samples of any interest are so complex that automatic control would not be possible. Moreover, some factors cannot yet be placed under computer control—the maintenance of optimum focus, or the detection and correction of instabilities in the lenses, spectrometers or other components. There is no substitute for the intelligent operator, but there is plenty of opportunity for computer-aided analysis: a small computer can store whole spectra, recall them for comparison with a standard, assist in identification of the peaks, carry out quantitative corrections during the analysis and, in general, act as a powerful aid in a personal investigation.

3.4 COMBINATION INSTRUMENTS

The problems and advantages of combining the functions of a scanning electron microscope and a microprobe analyser in one instrument were naturally discussed in section 3.1.1 when comparing the design of the two instruments. Another important combination is that of a transmission electron microscope and microanalyser, and this may be achieved by either scanning or non-scanning methods.

The scanning-transmission electron microscope has already been mentioned in section 2.5.8 and is discussed in more detail in chapter 8. It is essentially a scanning microscope with the electron detector placed beneath the specimen. It is clearly possible to combine this with a spectrometer to analyse the X-ray emission and this combination is increasing in popularity. Some redesign of the electron gun is necessary in comparison with a standard scanning microscope or microprobe since transmission electron microscopy has very limited applications to materials if the accelerating voltage is less than 100 kV.

It is also possible to add a spectrometer (usually energy-dispersive) to a normal transmission electron microscope. This can be useful for identifying precipitates, measuring grain boundary microstructures and the like, but sensitivities are often low and the standard transmission microscope will normally give an electron beam no finer than one or two microns. The problem of designing an instrument with high performance both as a microscope and as a microanalyser was solved by Duncumb and his group at Tube Investments laboratories. The instrument, known as EMMA (electron microscope microanalyser), is now produced by A.E.I. It has an extra condenser lens (a minilens, section 3.3.1) over the two normally used in the transmission microscope, to give a 100 nm focal spot, two focusing wavelength-dispersive spectrometers with a quite high takeoff angle (45°), accelerating voltages up to 100 kV and a resolution in transmission of 1 nm. It can therefore be used to study microstructure, defect structure, crystallography and chemical composition on the same specimen, and has been applied for example to the study of grain boundary microstructure in aluminium alloys, which is believed to control their stress-corrosion susceptibility; the distances over which the structure changes in these materials is so small that no reliable results can be obtained by other methods.

There is almost always a general advantage to be gained by attacking a problem with more than one technique, but there is also a specific gain in the spatial resolution of the microanalysis when the specimen is in the form of a thin film. The spatial resolution of analysis with a bulk specimen is limited not by the diameter of the electron beam (unless this be greater than one or two microns) but by the extent to which it spreads beneath the specimen surface (see section 2.4.3). The diameter of the interaction volume, usually of the order of one micron, can be reduced by physically removing part of this volume, that is, by using a thin film specimen. We can then expect the spatial resolution in microanalysis to be of the same order as the film thickness; this is generally 100–500 nm in transmission electron microscopy, so an improvement in spatial resolution of up to an order of magnitude is obtained. There is a corresponding loss in the intensity of the X-ray emission, so relatively long counting times must be used to overcome the statistical limitations discussed in the last section.

3.5 QUANTITATIVE ANALYSIS WITH THE ELECTRON MICROPROBE

The ultimate precision of the electron microprobe is limited by the statistical noise, as discussed in section 3.3.3 and in more detail in Ziebold (1967). The normal procedure in microanalysis is to compare the count rate for a given element, first from the area on the sample under analysis then from a standard of the pure element or of an alloy whose composition is accurately known. The ratio of the intensities from the unknown and the standard, k, is the basic experimental measurement (since there is always some background noise, the appropriate values must be subtracted from both intensities before division). In practice it is found that a precision on k of better than 0.5 per cent is attainable in modern instruments with counting times of 100 s when $k > 0.1$ (Beaman and Isasi, 1972) implying a solute concentration above about 10 per cent; the statistical errors in k for a given counting time increase rapidly below this figure, so long counting times are needed as the analysis approaches the limits of detectability. To obtain an overall accuracy of this precision, great care must be taken to exclude all other sources of error: accelerating voltage uncertainty and drift, spectrometer drift, counting losses at high rates, specimen stage drift and several other possible instrumental errors. The engineering of the instrument, and the operator's understanding of it, must be of a very high order to achieve accuracies better than 1 per cent in the measurement of k values.

The problem now is to convert k into values for the chemical composition, without losing accuracy. The first approximation is to take k itself as the chemical composition of the sample (expressed in weight fraction) but this can be wildly inaccurate; the errors can vary from less than one per cent to several hundred per cent, depending on the system. Beaman and Isasi (1972) cite the case of a simple alloy of magnesium and aluminium, containing 80 per cent magnesium and 20 per cent aluminium;

reliance on the uncorrected k values would give measured concentrations of 79.4 per cent magnesium but only 5.7 per cent aluminium. Normalisation of the values so that they add up to 100 per cent merely increases the error in the magnesium concentration (changing it from 79.4 to 93.3 per cent) while only marginally improving the aluminium value, from 5.7 to 6.7 per cent. Much greater accuracy can be attained by comparing the unknown sample with standards that are very similar in composition rather than with pure elements. However, this does defeat the main object of a general analytical method, which is to perform a quantitative analysis on a completely unknown specimen without the labour of preparing a series of alloy standards, each containing perhaps several elements. It is really necessary to understand the processes involved in the emission of X-rays from solids bombarded by electrons. The differences in the X-ray emission from an element when it is surrounded by like elements from that when it is surrounded by any combination of elements can then be expressed as a correction factor to be applied to the experimental k values.

The following corrections due to five separate physical processes are generally applied.

(1)   *The backscattered electron correction.* A certain fraction of the incident electrons are backscattered from the surface, so cannot produce X-rays in the sample. This fraction increases with the atomic number of the specimen and is therefore a function of the mean composition.

(2)   *The stopping power correction.* This is related to the efficiency of production of X-rays, and again to the atomic number of the specimen. The mean electron range will be greater in a light element but the ionisation energy per unit mass is also higher, and the overall generation of X-rays is lower. Since this and the first correction (which tend to cancel) are both controlled by the atomic number, they are often lumped together as the 'atomic number correction'.

(3)   *The absorption correction.* The X-rays are generated at various depths throughout the interaction volume. The intensity that escapes from the specimen is therefore dependent on the amount of absorption they experience during their passage to the surface. This will depend on the size of the interaction volume and on the absorption of the material (thus varying with the mean composition) and on the wavelength of the X-rays that are being examined. The absorption path length will also depend on the takeoff angle of the spectrometer, hence the need for high and accurately known takeoff angles. This is usually the most important correction.

(4)   *The characteristic fluorescence correction.* X-rays are produced by other X-rays as well as by electrons. The emission of any characteristic line will be caused by the primary beam and by any of the characteristic lines from other elements in the specimen that have a sufficiently high energy. Again, the magnitude of this effect depends upon the composition of the specimen.

(5)   *The continuum fluorescence correction.* This is similar in principle to the previous correction but arises from the continuous rather than the other characteristic radiation present. It is normally separated from the

characteristic fluorescence correction because it is calculated differently, and also because as the smallest of the corrections (usually less than 1 per cent) it is frequently ignored.

These factors will be discussed more fully in the following sections, but a detailed derivation of the equations describing them will not be attempted; that is a matter of considerable mathematical and physical complexity, of interest only to a specialist. Comprehensive treatments of the correction factors can be found in Duncumb and Shields (1963), Wittry (1964), Philibert (1964 and 1967), Martin and Poole (1971), and Beaman and Isasi (1972).

3.5.1 Atomic Number Correction

The problem is to calculate the number of emitted photons, of the wavelength being measured, per incident electron. First, the number of K ionisations (assuming that a K line, usually $K_{\alpha 1}$, is being measured) is calculated, per electron entering the material. This will be an integral over the energy of the electron, from its initial value E_0 (given by the accelerating voltage in the microprobe) down to the minimum value for K ionisation, E_c. The integrand will be linear with the ionisation cross-section Q (which is a function of energy) and inversely proportional to S, the 'stopping power' of the material. As S increases, the range of the electron decreases so fewer atoms can be ionised (S is also a function of energy). The number N of ionisation events occurring per electron entering the material is then given by

$$N = \alpha n^A \int_{E_0}^{E_c} \frac{Q \, dE}{S} \tag{3.7}$$

where n^A is the number of A atoms per unit volume in the material (A is the element being analysed) and α is a constant. n^A of course varies with composition, and so does S (which will depend upon the atomic number and concentration of all the other elements present) but the other terms in equation 3.7 are independent of composition. To find the number of X-ray photons generated, equation 3.7 is simply multiplied by the appropriate conversion efficiency which again is independent of composition as long as the state of chemical bonding of the element does not change with composition. The intensity ratio of X-rays generated from two different specimens, 0 and 1, is found by dividing two equations such as 3.7

$$k_g' = \frac{n_1^A \int_{E_0}^{E_c} \dfrac{Q \, dE}{S_1}}{n_0^A \int_{E_0}^{E_c} \dfrac{Q \, dE}{S_0}} \tag{3.8}$$

In the experimental measurement the intensity of the $K_{\alpha 1}$ emission from the specimen is compared with that from a pure sample. In that case

$$n_1^A / n_0^A = C_A \tag{3.9}$$

where C_A is the concentration which is to be measured (strictly, C_A as defined here is a volume fraction; however, it is normally taken as a mass fraction to give the common 'weight per cent' figure, and the definitions of S_0 and S_1 modified to allow for the difference in density of the specimen and the standard). Substitution of equation 3.9 into equation 3.8 therefore gives the stopping power correction, which has to be applied to the experimental k value. Although Q is the same function in both integrals, the stopping power is not. It will depend on the total composition of the specimen since for any one element i, S_i is given by, for example, the Bethe expression

$$S_i = \frac{\alpha Z}{EW} \ln \left(\frac{\beta E}{J} \right) \qquad (3.10)$$

where Z is the atomic number, W the atomic weight and J the ionisation potential of element i (α and β are constants). Then in the actual sample an average value of S must be taken, such as

$$S_1 = \sum C_i S_i \qquad (3.11)$$

It is clear that C_A will not equal k'_g. The next correction to be applied arises because electrons are lost from the primary beam by backscattering at the surface. The important quantity here is the amount of possible ionisation that is lost through backscattering (rather than simply the number of electrons lost) so integrals similar to equation 3.7 will again be involved. Writing the fraction of possible ionisation lost through back-scattering as $(1 - R)$ and noting that R will depend upon composition, the $K_{\alpha 1}$ intensity ratio generated per incident electron is obtained from equation 3.8 as

$$k_g = \frac{C_A R_1 \int_{E_0}^{E_c} \dfrac{Q \, dE}{S_1}}{R_0 \int_{E_0}^{E_c} \dfrac{Q \, dE}{S_0}} \qquad (3.12)$$

and the correction to be applied to the measured intensity ratio can be written as

$$C_A = (F(0)_0 / F(0)_1) \times k \qquad (3.13)$$

the zeros in parentheses acting as a reminder that so far the absorption of the material has been taken as zero.

This parameter, $F(0)_0 / F(0)_1$ is the atomic number correction. It is often given the symbol $\bar{Z}$. The stopping power part is evaluated by numerical integration using equation 3.10 and an expression due to Webster for the ionisation cross-section Q

$$Q = \frac{\text{const} \ln (E/E_c)}{E_c^2 \times (E/E_c)^{0.8}} \qquad (3.14)$$

It is evident that many parameters must be taken into account. The backscattered electron correction R_1/R_0 can be evaluated similarly, or by using an approximate expression for the integrand of R, due to Duncumb;

this is a 36-term polynomial, which can be handled quite conveniently in a computer program.

The stopping power and the backscattering corrections work in opposite directions and in some systems the overall correction $\bar{Z}$ can be very small—the very existence of the atomic number correction was disputed in the early days of microanalysis. However, calculations of C_A that ignore this effect can easily be out by ten per cent, that is, twenty times worse than the precision of the instrument.

3.5.2 Absorption Correction

The atomic number correction allows for concentration-dependent differences in the number of $K_{\alpha 1}$ photons that are generated inside the specimen and the standard. These photons must escape from the material before entering the spectrometer, and they therefore experience absorption (mainly photoelectric). If μ is the linear absorption coefficient of the material for the particular $K_{\alpha 1}$ wavelength and x is the path length of the photons in the material then the attenuation factor is $\exp(-\mu x)$. Unfortunately the photons are not all generated at constant depth in the material, so the attenuation factor must be integrated over depth and, to make matters worse, the maximum depth and the distribution of generated intensity over this depth will both vary with composition. This is illustrated in figure 3.11, which shows schematically the volumes generating X-rays in a light element, a heavy element and a mixture of the two.

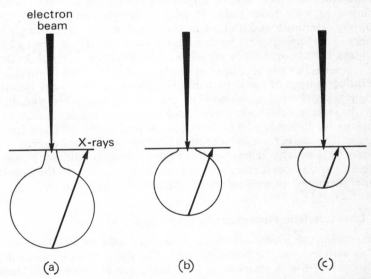

Figure 3.11. Schematic shapes of X-ray generation as a function of depth: (a) element of low atomic number, (b) alloy of (a) and (c), (c) element of high atomic number. The arrows show the absorption paths of the X-rays emitted in the direction of the spectrometer

Because of the dependence of the absorption correction upon the way in which the photons are generated as a function of depth, this correction is not really independent of the atomic number correction. However, it is normally treated separately. It is rather difficult to calculate from first principles (partly because of its partial dependence on $\bar{Z}$) so a semi-empirical approach has been adopted, in which an approximate mathematical form for the correction is derived and then fitted to the available experimental data (Philibert, 1962; Duncumb et al., 1969). The absorption correction is usually given the symbol A so the correction equation so far is

$$C_A = k\bar{Z}A \qquad (3.15)$$

Absorption errors are the largest in most analyses, since the absorption coefficients of two elements can differ widely. In the example of the aluminium–magnesium alloy quoted in the introduction to section 3.5 a value of A of 3.4 was responsible for most of the required correction. In other cases the correction may be only a few per cent, but it is still worth applying in order to use the full precision of the microprobe. Errors in the absorption correction itself can, unfortunately, amount to a few per cent if the accelerating potential, the takeoff angle or the absorption coefficient are not known precisely (Heinrich and Yakowitz, 1969). The precision of the first two of these factors is normally fixed by the manufacturer, and the second by the precision of the experimental data that is available. In a scanning electron microscope fitted with an energy-dispersive spectrometer on the other hand, the takeoff angle may not be known very accurately and may also vary from point to point through variations in surface topography. Heinrich and Yakowitz show that if the takeoff angle is 15° and there is a 1° uncertainty in this figure, errors in A amount to 2–5 per cent unless the absorption is very low. The situation is much better for heavy elements, whose $K_{\alpha 1}$ lines are only weakly absorbed (provided the accelerating voltage of the instrument is high enough to excite them), and for high takeoff angles. However, the takeoff angle is variable for energy-dispersive detectors, since these consist of a silicon slice several millimetres in diameter placed between 50 and about 500 mm from the specimen. The detector can subtend an angle of several degrees at the specimen (this is why it has such a high collection efficiency), and the precise absorption correction would have to depend upon the specimen-detector geometry as well as on the factors discussed above.

3.5.3 Characteristic Fluorescence Correction

The excitation of a characteristic X-ray line can be accomplished by X-ray as well as electron bombardment. The fluorescence caused by the $K_{\alpha 1}$ photons of the A element as they leave the specimen will diminish their intensity, but this is one of the mechanisms accounted for in the absorption correction. However, some of the $AK_{\alpha 1}$ photons will have been generated not by the electron beam but by fluorescence from X-rays emitted by other elements in the specimen. If the ratio of the intensities generated by these two causes is γ then the correction to be applied to the measured intensity is $1/(1 + \gamma) = F$, the fluorescence correction. The overall correction so far becomes

$$C_A = \bar{Z}AF \qquad (3.16)$$

and very often the correction process is taken no further.

If several elements that can cause fluorescence are present then γ must be summed over them all. Expressions for the correction have been derived by Castaing (1951) and Reed (1965); these in turn depend upon expressions for the fluorescence yield. The fluorescence correction is often less than 1 or 2 per cent (it is zero if the X-rays present from other elements are not sufficiently energetic to excite the line under consideration) but can be greater than 10 per cent in unfavourable cases. The present equations describing the fluorescence correction are not entirely satisfactory, partly because there is some shortage of experimental data on the factors involved (such as the fluorescence yield).

3.5.4 Continuum Fluorescence Correction

The principle of this correction is exactly the same as that of the characteristic fluorescence correction, but sufficiently energetic continuum radiation is always present (or the $K_{\alpha 1}$ line under analysis would not be present either). However, its intensity is always rather feeble in comparison with the characteristic lines and the consequent correction is rarely greater than 1 per cent. This is near the possible accuracy of the microprobe, so the correction is often ignored.

An expression for the continuum fluorescence correction has been derived by Henoc (1968); he uses approximate expressions for the intensity of the continuum as a function of wavelength and integrates the fluorescence contribution from the low-wavelength limit of the continuum to the absorption edge of the element under analysis. The expression is consequently quite complicated and requires as much computation as the first three corrections put together.

3.5.5 Application of the Corrections

The application of the $\bar{Z}AF$ corrections is clearly a very tedious business if done by hand, involving perhaps hours of calculation. Several iterations are usually required. The corrections are themselves dependent upon the composition of the alloy so a trial value must first be assumed (this would be the experimental intensity ratios obtained from the specimen and standard) and the calculations repeated until the results converge. This is obviously a task for a computer, preferably one that is on-line to the microanalyser so that dubious results may be checked instantly and decisions taken about the next stage in the study of the specimen. If only an off-line computer is available one of the published correction programs may be used, for example, Duncumb and Jones (1969), Mason *et al.* (1969), Shaw (1969) and Colby (1968). A critical review of computer programs for applying the $\bar{Z}AF$ corrections has been given by Beaman and Isasi (1970), who recommend the use of one of the programs cited above. A busy analytical laboratory might analyse several thousand specimens a year so either dedicated or off-line computers can effect real economies.

The $\bar{Z}AF$ or 'split' correction method described above is not the only way of accounting for the different conditions in the specimen and the standards. We have commented that the absorption correction is not independent of the atomic number correction and methods exist for treating the two simultaneously. However, any correction method for microanalysis is judged on practical rather than theoretical grounds—the availability of computer programs, the computation time required and the eventual accuracy of results. The split method performs well on these grounds, and the separation of the corrections does have some physical meaning. For example, after some experience with the magnitudes of the corrections, it is possible to judge the circumstances in which the fluorescence correction might be high; when analysing fine particles in these circumstances it is then useful to keep the takeoff angle low, so that secondary fluorescence (emission excited in regions below the small phase being investigated) is minimised by a long absorption path. A review of other methods of correction is given in Reuter (1971) and a detailed comparison in Poole (1968).

With good specimens, modern instruments operated correctly and the best available correction procedures the overall accuracy of analysis (determined experimentally by collating the results of many workers from carefully prepared standards) is likely to be ± 1 per cent in the most favourable cases (when all the corrections are small), ± 2 per cent in reasonably straightforward cases (when simple alloys are used, in which no element is more dilute than about 20 per cent) and ± 5 per cent in general cases, including the analysis of ceramics and minerals. The errors are likely to be around this last figure even in favourable cases when energy-dispersive spectrometers are used, but this figure could improve when the cause of the high background observed with these spectrometers is properly understood. The errors increase substantially for light elements ($Z \leqslant 10$) and for elements that are only present in low concentrations, even though trace elements can sometimes be detected at concentration levels of 50 parts per million. It is worth remembering that the total mass of material analysed by the electron probe is about 10^{-11} or 10^{-12} g and that the limit of detectability can be as low as 10^{-16} g.

3.6 COMPARISON OF MICROANALYTICAL TECHNIQUES

Some representative data on the four microanalytical techniques discussed in this chapter are given in table 3.2. The choice of technique will to a large extent depend upon the application, but it is evident that the electron microprobe is, at present, the most versatile instrument when quantitative as well as qualitative analyses are required for a wide range of elements. The other instruments have great merits for particular applications. The

Table 3.2. A comparison of microanalytical techniques
(The figures that are likely to improve with improved technology or better understanding of the processes involved are marked with an asterisk)

Factor	Electron Microprobe	Laser Microprobe	Ion Microscope	Auger Spectrometer
Spatial resolution of analysis, μm	0.5–1	1–2	1–2*	0.1*
Depth of analysis, μm	0.5–2	1–2	<0.005	<0.005
Minimum mass of sampled volume, g	10^{-12}	10^{-8}	10^{-13}*	10^{-16}*
Detectability limits, (1) mass, g	10^{-16}	5×10^{-10}	10^{-19}*	10^{-18}*
(2) conc., wt. ppm	50–1000	1–10	0.01–100*	10–1000*
Elements analysable	Z ≥ 4	'Metallic' elements Li, Be	All elements	Z ≥ 3
Performance for Z < 10	Poor	±10%*	Good	Good
Quantitative accuracy, C ≥ 10%	±1–5%		Not yet* established	Not yet* established
Vacuum required, torr	10^{-5}	760	10^{-8}	10^{-10}
Damage to specimen	Yes, in poor thermal conductors	Sampled vol. destroyed		Rare
Approx. exposure for point analysis, seconds	100	3×10^{-8}	0.05	1000
Approx. cost (advanced optical microscope = 1 unit)	8	1	16	8

laser-microprobe analyser is the cheapest and the easiest to operate. It has adequate sensitivity for many applications, analyses can be made reasonably accurate and although only metallic elements can be analysed (including semiconductors such as silicon) there is no particular problem with the light metals lithium and beryllium. This instrument can be used very effectively in materials-based industries for quality control or for trouble-shooting (for example, identifying inclusions in alloys or unexpected phases in ceramics). A fuller understanding of the physics of the processes involved, such as laser-induced evaporation, should improve the accuracy (Moenke *et al.*, 1970).

Of all the methods described in this chapter, it is those based on Auger electron spectroscopy that are developing the fastest, and each year brings significant improvements in the techniques. Even now, the Auger method has the highest spatial resolution of microanalysis (100 nm). The resolution during a scanned image is about an order of magnitude worse (that is, about the same as a microprobe) but these limits are determined by the present technology, in particular that of the electron gun, rather than by any fundamental limitation as in the case of the electron microprobe or laser microanalyser. From the analytical point of view, the Auger method has one very considerable superiority over the electron-microprobe method based upon X-ray emission, in that light elements emit Auger electrons much more efficiently than X-rays, and moreover, the energies of the Auger electrons lie in a band for which detection and energy measurement is easy. When the procedures required for quantitative analysis with the Auger method have been thoroughly worked out, the analysis of light elements is likely to be an important application. Another advantage of Auger spectroscopy is that it is easily combined with scanning electron microscopy; in the future the Auger technique will probably be regarded as simply one of the modes of operation of the scanning electron microscope.

The ion microscope also works very well for the light elements. It can even analyse for hydrogen and helium (though the sensitivity for helium is not very good), which, lacking outer-shell electrons, emit neither Auger electrons nor characteristic X-rays. The effective sample size is rather larger than the sampled volume in Auger work because ion beams cannot as yet be focused as finely as electron beams, while the depth of the sampled volume is similar in the two cases. The detectability limit for the ion microscope is probably the best of all analytical techniques (except the highly specialised single-atom analysis possible in the field–ion microscope, chapter 7) for two reasons: the ion sources have a high brightness, so that shot noise does not in practice limit the spatial or analytic resolution, and mass spectrometers with very high mass resolution are quite straightforward to build, so that an isotope of any element can be detected and the signal displayed as a sharp peak against a zero background. As far as 'mainstream' analysis of materials is concerned, the ion microscope will probably find its main applications in the detection of trace elements, light elements (for example, the study of internal oxidation), in the study of polymers and natural materials, which contain a high proportion of light elements, and in the study of impurity distributions in semiconductors, in which the impurity levels are very low. Again, the factors involved in accurate quantitative analysis remain to be worked out, although good results have been obtained in specific cases.

Both the Auger and the ion microscope techniques require an ultra-high-vacuum environment around the specimen; a constraint which, for routine analysis of materials, is considered a disadvantage because of the complexity and cost of the equipment. The requirement arises because in both these methods the sampled volume is at most a few atoms deep and it is necessary to avoid any spurious effects due to surface contamination. Indeed, as pointed out in sections 3.2.3 and 3.2.4, it is often extremely difficult to be sure that the sampled volume is really typical of the bulk material and caution is always necessary when generalising on the results of a surface analysis. The fact that the analysis is confined to the surface layers does, however, mean that these two techniques are very appropriate for the study of surfaces. The properties of surfaces are believed to be of fundamental importance in the behaviour of many aspects of solids—the kinetics of phase transformations, the initiation of plastic flow, the onset of fatigue failure, the phenomenon of cleavage are some examples—and these techniques are powerful tools for the surface scientist. Some examples of interesting surface analyses are given in the next section.

3.7 SOME APPLICATIONS OF MICROANALYSIS

The analysis of the chemical composition of a small feature of the microstructure of a specimen, or of a small fragment of material, is such a general requirement that it is only possible to give a few illustrative examples of the applications of the techniques. There is, of course, considerable overlap between the capabilities of the four instruments described in this chapter, but the electron microprobe is by far the most widely used. Applications that depend upon a general ability to perform microanalysis will therefore be described with reference to this technique. The applications described in the sections on the laser microprobe, the Auger spectrometer and the ion microscope will be those which cannot conveniently (if at all) be tackled with the electron microprobe.

3.7.1 Applications of Electron-microprobe Analysis

The majority of the scientific and industrial applications of microanalysis of materials may be classified under one of three headings: the study of diffusion processes, the characterisation of phase transformations and the identification of small fragments or phases.

From the scientific point of view it is important to have accurate values for diffusion coefficients over a range of temperatures. Diffusion couples are frequently used for such measurements: the two materials concerned are placed in close contact and, after diffusion has proceeded for some time, the penetration profiles are measured. The diffusion coefficients can be calculated from these profiles using Matano's equations (see for example Christian, 1965; any other experimental arrangement for which the diffusion equations can be solved may of course also be used). These methods do not absolutely require the use of an electron microprobe, since in earlier days both radioactive tracer techniques, and mechanical sectioning followed by bulk analysis were used; however, the microprobe has a higher spatial resolution than either of these techniques and the method is not destructive. The inconvenience of using radioactive tracers is also avoided.

Since diffusion is one of the fundamental processes involved in the production and in the heat treatment of materials, its study by microanalysis forms an important industrial application. Besides the more obvious uses, such as checking whether an annealing treatment has in fact produced a homogeneous alloy, there are applications which depend strongly upon the microscopic nature of the analysis: the study of thin diffusion bonds, of the high-temperature breakdown of a thin cladding layer or of the fibre–matrix interface in a composite material. An unusual example is the work of O'Boyle et al. (1969) on nuclear fuel elements made from a mixture of uranium and plutonium dioxides. Measurement of the concentrations of plutonium at various points enabled them to show that plutonium diffusion had taken place during service even though there was no initial concentration gradient in the material; the driving force was the temperature gradient existing in the fuel element under the conditions of the reactor (this phenomenon is sometimes called 'Soret diffusion' and is also troublesome in the growth of high-quality multi-component single crystals). They were also able to identify numerous fission products which occurred as inclusions in the oxide matrix.

Accurate phase diagrams are essential aids to the comprehension of microstructures of solids, be they metals, polymers, ceramics or minerals. Microanalytical methods can greatly increase the speed (and in some cases the accuracy) of the determination of phase diagrams. A single specimen can be used to determine the boundaries of an entire two-phase region in a binary alloy (or of more complex phase mixtures in multi-component alloys); the specimen is annealed at various temperatures, and quenched and analysed after each anneal. Determination of the compositions of each phase present automatically gives the compositions of the ends of the tie-lines at that temperature. Without microanalysis such regions can only be determined by making up a series of alloys of different composition and bracketing the phase boundary by annealing them at different temperatures—a much more laborious procedure. An even more elegant method is to use a diffusion couple (for a binary system) which has been given a prolonged anneal at the temperature of interest. Since the couple will contain all possible compositions of alloys that can be formed by the two elements, it should also contain all possible phases. Analysis of the phases formed along a section of the couple should then determine a complete horizontal section of the phase diagram. There are pitfalls in this method, since some phases may be formed in sections that are too thin for accurate analysis, and also since the whole system is in a steady state rather than true equilibrium.

However, it is certainly a powerful aid to conventional methods, as illustrated by the work of Adda and his co-workers on the uranium–nickel system (Adda *et al.*, 1961). In some cases the quantitative accuracy of the electron-probe microanalyser is inadequate for the use for which the phase diagram is needed, but even then it can be invaluable for quickly sketching in the phase boundaries and thus reducing the time and labour that have to be spent on more accurate techniques.

The applications mentioned so far have generally been those that demand the greatest possible accuracy in the quantitative analysis of the material or of the small phase. The scanning electron microscope fitted with an energy-dispersive detector is not, therefore, very suitable for such work. Other aspects of the study of phase transformations or of the identification of small particles do not require such high accuracy, and the high speed of the energy-dispersive detector is then advantageous. The study of segregation is an example. It is often important to know whether a solute is uniformly dispersed within a grain or has accumulated at grain or cell boundaries. If the solute concentration is so low that a scanning X-ray image is too noisy properly to show the solute distribution, the 'line-scan' method can be used; the electron beam is set so as to scan very slowly in a line across the suspect boundary zones, when variations in solute content will show up as variations in the X-ray signal. It is often possible to superimpose the ratemeter output signal on the scanning electron image so that a very close correlation between the two may be made. For such applications, which require high sensitivity rather than high accuracy, the scanning electron microscope plus an energy-dispersive detector may frequently prove the better instrument. Examples of this type of application are shown in figures 3.12 and 3.13.

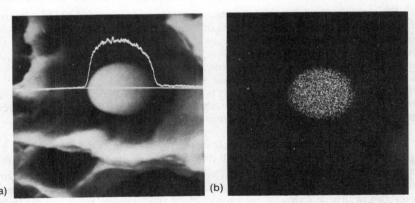

(a)　　　　　　　　　　　　　　　　　(b)

Figure 3.12. Steel fracture surface showing an inclusion (measuring 3 μm × 4 μm) in an area of ductile rupture. (a) Secondary electron image plus 'line scan' in which the electron beam is scanned along the horizontal line and the irregular line is the display of the intensity of the MnK_α emission across the inclusion; (b) a scanning X-ray image of the MnK_α emission confirms that manganese is an important constituent of the inclusion. Note that accurate quantitative analysis is not necessary for this application, but could be used to identify the inclusion more precisely. (Courtesy of Cambridge Scientific Instruments Ltd)

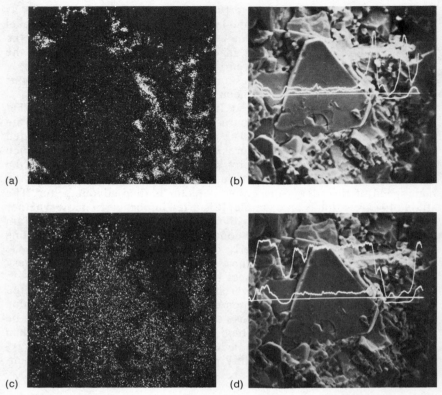

(a)   (b)

(c)   (d)

Figure 3.13. Tungsten carbide specimen (tungsten carbide powder sintered with cobalt) fractured by transverse rupture. Line scans and scanning X-ray images show the distribution of the cobalt binding. (a) CoK_α X-ray image; (b) secondary electron image plus CoK_α line scan; (c) WL_α X-ray image; (d) secondary electron image plus WL_α line scan. Although the exposure times were short for the scanning images so that shot noise is evident, the resolution is sufficient to show that the cobalt phase is inhomogeneously distributed. The large triangular tungsten carbide particle (the main feature of the image) is a few microns across. (Courtesy of Cambridge Scientific Instruments Ltd)

The identification of small inclusions in materials, as illustrated by figure 3.12, is a major industrial use of the microprobe, because the presence of inclusions may indicate some problem in the production process which may be resolved by identification of the inclusion. In some cases the inclusions themselves may affect the properties of the material rather severely, for example by reducing the fracture toughness. Although much attention has been paid to inclusions in metals (notably steels, because of the vast tonnage that is produced), similar problems can occur in other materials; the inclusions of titanium dioxide shown in the picture of a nylon fibre (figure 2.28) were in fact identified by an energy-dispersive spectrometer fitted to the scanning electron microscope. There can at times be problems in the accurate analysis of small inclusions, since it is possible for the beam to penetrate the inclusion and also excite X-rays from the underlying matrix. As mentioned earlier, a shallow takeoff angle can ease this problem by increasing the absorption path length for X-rays originating relatively deep in the specimen. However, if the inclusion is small in all dimensions, part of the absorption path will lie in the inclusion and part in the matrix, so that the absorption correction is almost impossible to apply. The only way around this problem is to use the extraction replica technique (described in section 4.4.1) to remove the inclusion from the matrix and analyse it in isolation.

The extraction method is also very valuable if the compositional analysis alone is insufficient and crystallographic data is also required. After analysis, the specimen can be transferred to a transmission electron microscope and diffraction patterns obtained. The best instrument for this purpose is of course EMMA, the combined electron microscope and microanalyser (section 3.4) and both quantitative analyses and diffraction patterns have been obtained from particles as small as 100 nm with this instrument (the effect of beam spreading being obviated when the specimen itself is very small). Figure 3.14 illustrates this type of application, from the work of Jacobs (1973). The specimen was an oxide film stripped from an aluminium specimen containing only 0.03 per cent magnesium. Figure 3.14a shows an electron micrograph, and a diffraction pattern obtained from one of the crystals about 0.2 μm diameter (simply to obtain a diffraction pattern from a well-identified area this small is not possible without a minilens or other extra condensing lens in conventional transmission electron microscopes). The matrix was shown by diffraction to be amorphous. Figures 3.14b and 3.14c show the AlK_α and MgK_α peaks obtained from a 0.2 μm crystal. The peak-to-background ratio is good and after sampling various areas and applying the appropriate corrections the Al:Mg ratio was found to be 3:1 (compared with over 3000:1 in the original alloy) in both the crystals and the amorphous matrix. Together with the diffraction pattern these results identified the 'oxide' as a two-phase non-stoichiometric $MgAl_2O_4$-type spinel. This analysis required both the fine-scale compositional analysis and the diffraction pattern (the normal oxide γ-Al_2O_3 gives an almost identical electron diffraction pattern) and thus could not have been achieved with conventional electron microscopes and microprobes.

There are of course many applications of microanalysis that are quite unconnected with materials studies: biological and botanical work, geological research, environmental studies such as the analysis of particles of pollutant in the atmosphere, and forensic studies involving the identification and comparison of small fragments of material. One such investigation is illustrated in figure 3.15. This is a fragment of paint from a car involved in a hit-and-run accident, and analysis of the pigment was used to identify the type of car involved. Such fragments can then be compared with fragments from suspected vehicles, both by microanalysis and by optical and scanning electron microscopy. Paint analysis has also been used in the detection of art forgeries, since it can indicate the use of

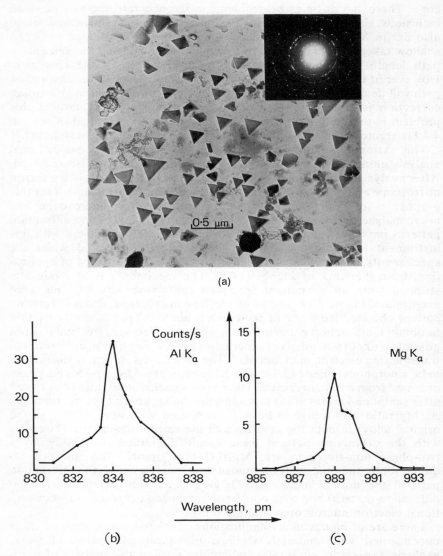

(a)

(b)

(c)

Figure 3.14. Application of EMMA (electron microscope microanalyser) to the analysis of small particles. The specimen is an oxide film stripped from aluminium containing 0.03 weight per cent of magnesium. (a) Electron micrograph of film; the inset shows the electron diffraction pattern obtained from one of the crystals, ~200 pm diameter; (b) X-ray spectrum near AlK_α peak; (c) spectrum near MgK_α peak, both obtained from a crystal ~200 pm diameter. (Courtesy of M. H. Jacobs, Tube Investments Research Laboratories, Hinxton Hall, Cambridge)

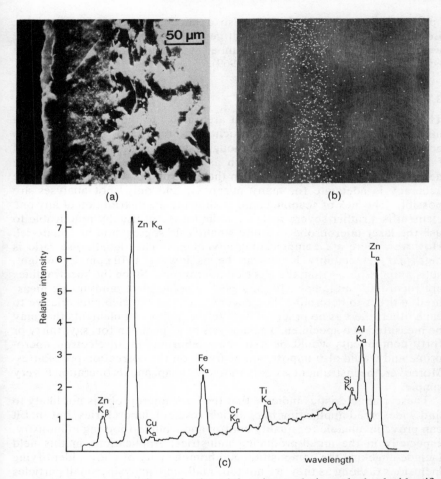

(a)

(b)

(c)

Figure 3.15. A forensic application of the microanalytic method—the identification of paint fragments found after a hit-and-run accident. (a) Optical micrograph of cross-section of fragment; (b) scanning X-ray image showing the distribution of zinc; (c) energy-dispersive spectrum from the zinc-rich region. This sort of information can be used to identify the paint layers involved, and hence indicate the type of vehicle, then to provide a 'fingerprint' for a detailed match with a suspected vehicle. (Courtesy of C. G. van Essen and the Metropolitan Police Forensic Laboratories)

pigments not available at the alleged period of the painting (see for example Ogilvie (1960) in an article winningly entitled 'The Lady is a Phoney').

In this short space we can do no more than give a glimpse of the endless possibilities of the techniques of microanalysis. More comprehensive reviews have been given by Melford (1962), Wheeler (1968) and Fleetwood (1968) for metallurgical applications, Keil (1967) for mineralogical applications (much of which is also applicable to ceramic materials); Beaman and Isasi (1972) give a brief general review that is exceptionally well documented and there are a number of general books such as the one edited by Tousimis and Marton (1969).

3.7.2 Laser-microprobe Analysis

The applications of this instrument are similar to those of the electron microprobe, but they are very much more restricted. The analysis is confined to metallic elements (although this term should be interpreted fairly liberally since elements such as silicon and phosphorus can be determined). It is less accurate than the electron microprobe, though the accuracy is adequate for many purposes; and only spot analyses are possible. The lack of scanning images showing the distribution of any one element is a rather severe restriction, and it is not usually practicable to use the laser microprobe to study small-scale segregation in a material. However, there are compensating advantages. The signal/noise ratio is high and detectability limits can be as low as 1–10 ppm by weight, substantially better than the electron microprobe. Since the spectral lines are recorded simultaneously on a photographic plate, and are all measured in order to determine the concentrations of the elements relative to each other, there is no practical limit to the number of elements that may be measured in a specimen. The analysis of a specimen for, say, thirty or forty components would be extremely laborious in an electron microprobe and would also impose some strains on the correction procedures. Moreover, the instrument is comparatively cheap, and its operation is very simple.

These characteristics indicate that the laser microprobe is not likely to find widespread application in materials research laboratories, but that it can prove invaluable for quality control and trouble-shooting in industry, especially in the metal-producing industries. Applications in this field include checking the composition and homogeneity of alloys, identifying inclusions (as long as they are not too small) and analysing small particles accidentally introduced into components, to discover their origin. Specimen preparation for laser-microprobe analysis is simpler than for electron-microprobe analysis (especially for insulating materials); this clearly increases the overall speed of the analysis and reduces its cost. Some of the forensic applications discussed in the last section may also conveniently be performed using this instrument.

Since this instrument is relatively new, it is not yet used so widely as the electron microprobe and there is less experience of the types of problem for which it will prove most useful, but the references cited in section 3.1.4 give some further examples of current applications.

3.7.3 Auger Electron Spectroscopy

There is little doubt that the techniques necessary for accurate quantitative analysis using Auger spectroscopy will be worked out in due course, when the physics of the interaction is properly understood. The processes may even turn out to be simpler than those required for the electron microprobe; for example, the interaction volume is so small that the difficult absorption problems encountered when analysing very small particles in the microprobe are unlikely to find a counterpart in the Auger method. Potentially, therefore, the technique is extremely powerful, probably even more versatile than the electron microprobe since there is no difficulty with elements of low atomic number, and the spatial resolution is considerably higher.

At the time of writing, the applications of the technique are limited both by the understanding of the physical processes involved and by the technology of the instrument. High-brightness electron guns are essential, and although these are currently available their stability is not as high as one would wish for quantitative analyses. The ultra-high-vacuum requirement also means that operations such as changing specimens, filaments or apertures will never be so convenient to perform as they are on the electron microprobe, and the technique will be restricted to rather specialised applications for some time to come. As for the physical understanding, much more study needs to be undertaken of factors such as the characterisation of Auger peaks for all elements, changes in peak position and intensity with the chemical state of the atom involved, the precise calculation of the interaction volume and the energy loss suffered by escaping electrons and the secondary ionisation caused by these electrons as they leave the specimen. It can be seen that these problems are rather similar to those originally encountered in quantitative analysis in the electron microprobe.

Despite these problems there have been some strikingly successful applications of the method, primarily in the field of surface or interface segregation, which have utilised the very shallow depth of analysis to detect extremely thin layers, perhaps only one atom deep. The problem of grain boundary embrittlement by minute amounts of impurity is a good example. A simple deformation stage (in one case, no more than a remote-controlled hammer) is mounted inside the experimental chamber, the specimen is fractured while under the ultra-high vacuum and the resulting fracture surface (which is thereby clean on an atomic scale) is examined by Auger spectroscopy. The Auger signals can be compared with those from a pure standard, and an ion gun can be used to erode the surface so that changes in composition with depth can be followed. When the fracture surface is a grain boundary the technique thus provides evidence for grain boundary segregation on an extremely fine scale. Several workers have applied this method, for example Palmberg and Marcus (1969) for iron, Joshi and Stein (1970 and 1971) to tungsten and copper and Powell and Mykura (1973) to copper. The grain boundary embrittlement that occurs in tungsten when it is annealed is notorious (at least to users of electron-beam instruments with tungsten filaments) and it has been suspected for a long time that the cause was segregation of trace quantities of oxygen to the grain boundaries. Joshi and Stein set out to measure this segregation by Auger spectroscopy; they found no oxygen segregation but instead a concentration of phosphorus of about 20 per cent in the layer within one or two atomic distances of the fracture surface even though phosphorus was not detectable by ordinary chemical analysis in the bulk material. While this does not of course prove that tungsten embrittlement is always caused by phosphorus, the experiment is clearly remarkably effective. The same authors, working with copper, showed that small amounts of bismuth (already known to cause embrittlement) segregated to form a monolayer at the grain boundaries; this result was confirmed by Powell and Mykura, although the latter showed that the segregation was not even from grain to grain.

The experiments described so far were performed with relatively coarse electron beams ($\sim$250 μm diameter), and correlation between the composition and the microstructure of the fracture surface was rather crude. Scanning Auger images are now available with much higher resolution and one of the earliest examples (also taken from the copper–bismuth work of Powell and Mykura) is shown in figure 3.16 together with a secondary electron image of the same area and Auger spectra taken from various points. Results obtained so far indicate that atoms are much more mobile at or near the surface than was previously suspected, and dynamic segregation (section 3.2.4) may turn out to be a serious problem in quantitative work.

3.7.4 Secondary Ion Emission Analysis

The scope of this technique is even greater than that of Auger spectroscopy, since hydrogen and helium can be analysed using the ion microscope. The sensitivity for helium is not very good, but that for hydrogen is satisfactory; this may lead to a range of applications from the investigation of hydrogen embrittlement in steels to the study of biological materials.

Although reasonably accurate quantitative analyses can be performed on a relative basis, for example the determination of isotope abundance ratios in lithium (Barrington *et al.*, 1966) or lead (Andersen, 1969), the situation is similar to that of Auger electron spectroscopy in that the procedures and techniques necessary for quantitative analysis have not been fully developed. The limitations are mainly those of theory rather than of technology; although an ion gun capable of producing a finer beam than those currently available (duoplasmatron guns—see the review by Socha, 1971) would be quite useful, the ion intensity is high and stability is not a problem over the short exposures required. However, factors such as the sputtered volume and sputtering rate, the chemical state of the surface layers of atoms, and the influence of the sputtering gas have not been worked out in detail and comparison with very similar standards is still needed for accurate analysis.

It often happens that a knowledge of the distribution of an element in relation to other aspects of the microstructure is of greater importance than a precise analysis. The ion microscope is already capable of this sort of work, and its extremely high sensitivity (in the parts per thousand

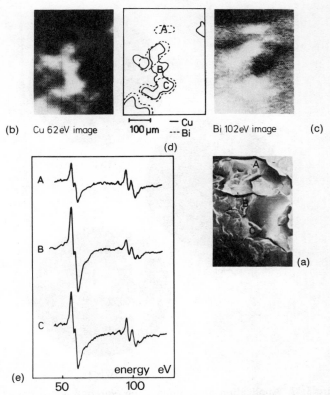

(b) Cu 62eV image   100 μm   — Cu
                                  --- Bi   Bi 102eV image   (c)

(d)

(e)   energy eV
    50   100

Figure 3.16. Application of Auger spectroscopy to the grain boundary segregation of bismuth in copper. The alloy was fractured inside the microscope. (a) Secondary electron (SEM) image; (b) scanning Auger image of copper distribution; (c) image of bismuth distribution; (d) diagram of the two distributions; (e) point analyses taken from the regions marked A, B, C on figure (a), showing the differing compositions at different regions in the grain boundary. The 'peaks' in the spectra have been electronically differentiated to enhance their visibility, thus each peak has a positive and negative excursion. The three peaks clearly visible are copper, bismuth and copper again. (Courtesy of B. D. Powell)

million range, obtainable because of the high signal/noise ratios possible with mass spectrometers) can make it more useful than the electron microprobe for this sort of work. An example is shown in figure 3.17, from the work of Castaing and Slodzian (1962). The specimen is an alloy of Si, Mg and Al and mass images of each of these components are shown. It is clearly seen that the matrix is primarily aluminium, that some virtually pure silicon particles are present and that the dendritic particles are magnesium containing some silicon. The quality of these images should be compared with those obtained on the electron microprobe; no noise is evident, yet the exposure times in the ion microscope are only of the order of seconds even if light elements are being imaged.

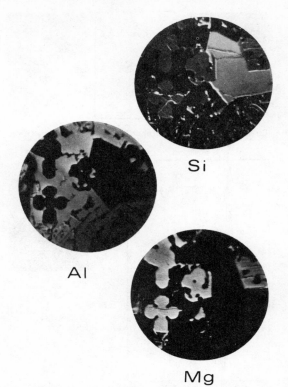

Figure 3.17. Application of secondary ion emission microscopy, to an alloy of alumium, silicon and magnesium. Mass images from each of these elements are shown. Field of view approximately 270 μm. Note the noise-free nature of the images, in contrast to those obtained in the electron microprobe. (Courtesy of G. Slodzian)

3.8 GUIDE TO FURTHER READING

References to reviews and original articles on Auger spectrometry, secondary ion emission analysis and laser-microprobe analysis have been given in the relevant sections, as have references to the correction procedures required for quantitative analysis with the electron microprobe. The most comprehensive modern review of microanalytical techniques is that by Beaman and Isasi (1972), which also contains a critical bibliography. A shorter general review is given by Reuter (1971) in a very useful volume of *Surface Science* which also contains articles on Auger spectroscopy (Chang, 1971) secondary ion emission analysis (Socha, 1971) and other techniques for the study of surfaces such as low-energy ion scattering. A review of instrumentation for the electron microprobe is given by Duncumb (1969); the EMMA instrument is described therein but more details can be found in Cooke and Duncumb

(1966). There are also a number of books on electron-probe microanalysis, for example that edited by Tousimis and Marton (1969), and a series of international conferences on X-ray Optics and Microanalysis; the third conference (Pattee *et al.* 1963, held at Stanford) the fourth (Castaing *et al.* 1966, held at Orsay) and the fifth (Möllenstedt and Gaukler, 1969) will be found particularly useful.

REFERENCES

Adda, Y., Beyeler, M., Kirianenko, A., and Maurice, F., (1961), *Memo. Scient. Rev. Metal*, **58**, 716

Andersen, C. A., (1969), *Proceedings of the Fourth National Conference of Microprobe Society of America*, Pasadena

Barrington, A. E., Herzog, R. K., and Poschenrieder, W. P., (1966), *Prog. in nucl. Energy*: Ser. 9, Analytical Chemistry, **7**, 243

Beaman, D. R., and Isasi, J. A., (1970), *Analyt. Chem.*, **39**, 859

Beaman, D. R., and Isasi, J. A., (1972), *Electron Beam Microanalysis*, ASTM Special Publication no. 506, ASTM, Philadelphia, Pa.

Castaing, R., (1951), Thèse, Université de Paris; and Publication no. 55, ONERA, Châtillon-sous-Bagneux (Seine)

Castaing, R., Deschamps, P., and Philibert, J., (eds), (1966), *Proceedings of the Fourth International Congress on X-ray Optics and Microanalysis (Orsay)*, Hermann, Paris

Castaing, R., and Slodzian, G., (1962), *J. Microsc.*, **1**, 395

Chang, C. C., (1971), *Surf. Sci.*, **25**, 80

Christian, J. W., (1965), *Theory of Transformations in Metals and Alloys*, Pergamon, Oxford, 351

Colby, J. W., (1968), *Adv. X-ray Analysis*, **11**, 287

Cooke, C. J. and Duncumb, P. (1966), in Castaing *et al.* (1966), 467

Duncumb, P., (1969), *J. scient. Instrum.* Ser. 2, **2**, 553

Duncumb, P., and Jones, E. M., (1969), *Electron Probe Microanalysis: An Easy-To-Use Computer Program For Correcting Quantitative Data*, Tube Investments Research Laboratories Report, Hinxton Hall, Cambridge

Duncumb, P., and Shields, P. K., (1963), *Br. J. appl. Phys.*, **14**, 617

Duncumb, P., Shields-Mason, P. K., and da Casa, C., (1969), in Möllenstedt and Gaukler (1969), 146

Fleetwood, M. J., (1968), in *Electron Microscopy and Micronalysis of Metals*, (eds J. A. Belk and A. L. Davies), Elsevier, Amsterdam, 225

Goldfinger, G., (ed.) (1970), *Clean Surfaces: their Preparation and Characterisation for Interfacial Studies*, Dekker (Marcel), Maidenhead

Hallerman, G., and Picklesimer, M. L., (1969), in Tousimis and Marton (1969), 197

Heinrich, K. F. J. and Yakowitz, H., (1969), in Möllenstedt and Gaukler (1969), 151

Henoc, J., (1968), in *Quantitative Electron Probe Microanalysis*, (ed. K. F. J. Heinrich), NBS Special Publication 298, Washington, 197

Henry, W. M. and Blosser, E. R., (1970), in *Techniques of Metals Research*, (ed. R. Bunshah), Interscience, New York, vol. 3, chapter 1

Jacobs, M. H., (1973), *J. Microsc.*, **99**, 165

Joshi, A., and Stein, D. F., (1970), *Metall. Trans.*, **1**, 2543

Joshi, A., and Stein, D. F., (1971), *J. Inst. Metals*, **99**, 178

Keil, K., (1967), *Fortschr. Miner.* **44**, 4

Le Poole, J. B., (1964), *Proceedings of the 3rd European Conference on Electron Microscopy (Prague)*, Czech. Acad. Sci., Prague, 439

Liebl, H., (1967), *J. appl. Phys.*, **38**, 5277

Martin, P. M., and Poole, D. M., (1971), *Metals and Mater.*, **5**, 19

Mason, P. K., Frost, M. T., and Reed, S. J. B., (1969), *B.M.–I.C.–N.P.L. Computer Programs for Calculating Corrections in Quantitative X-Ray Microanalysis*, National Physical Laboratory IMS Report 2.

Mayer, J. W. and Ziegler, J. F. (eds) (1974), *Ion Beam Surface Layer Analysis*, Elsevier Sequoia, Lausanne

Mela, M. J., and Sulonen, M. S., (1970), *J. scient. Instrum. (J. Phys. E)*, Ser. 2, **3**, 901

Melford, D. A., (1962), *J. Inst. Metals*, **90**, 217

Moenke, H., and Moenke-Blankenburg, L., (1966), *Einführung in die Laser-Mikro-emissionspektralanalyse*, Akademische Verlags, Leipzig. English translation *Laser Micro-Emission Spectroscopy*, Hilger, London, 1971

Moenke, H., Moenke-Blankenburg, L., Mohr, J., and Quillfeldt, W., (1970), *Mikrochim Acta*, 1154

Möllenstedt, G., and Gaukler, K. H. (eds), (1969), *Proceedings of the 5th International Congress on X-Ray Optics and Microanalysis (Tübingen)*, Springer-Verlag, Berlin

O'Boyle, D. R., Brown, F. L., and Sanecki, J. E., (1969), *J. nuc. Mater.*, **29**, 27

Ogilvie, R. E., (1960), *MIT tech. Rev.* **62**, 25

Palmberg, P. W., and Marcus, H. L., (1969), *Trans. Am. Soc. Metals* **62**, 1016

Pattee, H. H., Cosslett, V. E., and Engström, A., (eds) (1963), *Proceedings of the 3rd International Congress on X-Ray Optics and Microanalysis, (Stanford)*, Academic Press, New York

Philibert, J., (1962), in Pattee *et al.* (1963), 379

Philibert, J., (1964), *Métaux, Corros. Inds*, **40**, 157, 216, 325

Philibert, J., (1967), *J. Microsc. (Fr.)*, **6**, 889

Poole, D, M., (1968), in *Quantitative Electron Probe Microanalysis*, (ed. K. F. J. Heinrich), NBS Special Publication 298, Washington, 93

Powell, B. D., and Mykura, H., (1973), *Acta metall.*, **21**, 1151

Rasberry, S. D., Scribner, B. F., and Margoshes, M., (1967), *Appl. Opt.*, **6**, 81, 87

Redhead, P. A., Hobson, J. P., and Kornelson, E. V., (1968), *The Physical Basis of Ultrahigh Vacuum*, Chapman and Hall, London

Reed, S. J. B., (1965), *Br. J. appl. Phys.*, **16**, 913

Reuter, W., (1971), *Surf. Sci.*, **25**, 80

Rouberol, J. M., Guerne, J., Deschamps, P., Dagnot, J. P., and Guyon de la Berge, J. M., (1968), in Möllenstedt and Gaukler (1968), 311

Shaw, J. L., (1969), *A New Collection and Processing Method for Electron Probe Microanalysis Data*, UKAEA Report AERE-R6071

Slodzian, G., (1964), *Anns Phys.*, **9**, 591

Socha, A. J., (1971), *Surf. Sci.*, **25**, 147

Tousimis, A. J., and Marton, L. (eds) (1969), *Electron Probe Microanalysis*, Academic Press, New York

Webb, M. S. W., and Webb, R. J., (1971), *Analytica chim. Acta*, **55**, 67

Wheeler, M. J., (1968), in *Electron Microscopy and Microanalysis of Metals*, (eds J. A. Belk and A. L. Davies), Elsevier, Amsterdam, 200

Wittry, D. B., (1964), *Adv. X-Ray Analysis*, **7**, 395

Ziebold, T. O., (1967), *Analyt. Chem.*, **39**, 859

4

The Electron Microscope

The operation of the transmission electron microscope depends primarily upon the fact noted in chapter 2, that fast electrons are deflected by magnetic fields and can be focused by suitably constructed magnetic lenses. Thus in principle electrons can be used in the same way as light to form a magnified image of an object, but with the possibility that, since the wavelengths of the waves representing fast electrons are so short (3.7 pm at 100 keV) the resolution will be much higher than is possible with an optical microscope.

Electron microscopes generally work in transmission, whereas in the optical case materials are also observed in reflection. If a surface is illuminated with a beam of electrons near to normal incidence the backscattered intensity is low and most of the scattered electrons have lost appreciable amounts of energy. Since magnetic lenses suffer from severe chromatic aberration, any image formed with these electrons has low resolution as well as low intensity, and a direct electron analogue of an optical reflection microscope working at normal incidence is not feasible. As the angle of incidence is decreased, the specimen acts more like a mirror and the scattered electron intensity rises while the average energy loss falls, until near grazing incidence, images with good intensity and a resolution of 30–50 nm can be obtained. The contrast is similar to that produced by the scanning electron microscope when using the scattered electrons to form the image, and the resolution is also comparable. Because the optic axis of the detecting system makes only a small angle with the surface of the sample, images in the reflection microscope suffer from severe foreshortening. The scanning microscope (SEM) has the advantage of more modes of contrast production, with the possibility of image processing, and is easier to operate since the orientation of the surface to the incident beam is less critical. In addition, lower beam currents are required, the high beam currents necessary in the reflection microscope tending to damage the specimen. For these reasons, the scanning microscope has displaced the reflection microscope as the standard reflection instrument.

The operation of the electron microscope in transmission depends upon the additional fact that it is possible to prepare specimens that are thin enough to transmit a beam of electrons without too great a loss of intensity. The maximum usable thickness depends upon the atomic number of the material in question, but is typically 500 nm or more for specimens composed of light elements, decreasing to less than half this value for materials of high atomic number.

The higher the electron energy the better the transmission through the specimen, so microscopes are generally constructed to run at the highest accelerating voltage that can conveniently be used, typically 100 kV in standard instruments. Commercial instruments are now becoming available which operate at energies of up to 1 MeV, giving improved transmission but with consequent increases in size, complexity and cost.

As they pass through the thin film used as the object, the electron waves interact with its interior structure and may suffer consequent changes in amplitude and phase. By operating the microscope correctly these changes may be made visible in the resulting image as contrast. If it is known how different structures modify the transmitted electron waves, this may be interpreted to give information about the microstructure of the object.

How the electron wave interacts with the specimen is considered in the next chapter. This chapter is concerned with how the microscope forms the image, and how this may affect the contrast which is ultimately observed. The methods most widely used for preparing specimens thin enough for transmission microscopy are summarised, and some of the attachments that have been devised to increase the versatility of the basic instrument are briefly described.

4.1 BASIC DESIGN OF THE MICROSCOPE

The layout of the main components of a typical electron microscope is shown schematically in figure 4.1. The optic axis of the instrument is usually vertical, and at the top of the column, which is under vacuum, is a gun which produces the electron beam. This is focused by one or more condenser lenses onto the specimen which is mounted in a holder which can be withdrawn through an airlock, so enabling specimens to be changed without letting air into the system.

The specimen is usually mounted on a thin metal grid (typically about 3 mm in diameter) which is held in the specimen holder by a retaining ring or cap. If the specimen is not attached to the supporting grid it is held in position by another similar grid placed on top. Because the grid not only provides support but also conducts away the heat produced in the specimen by the beam, it is clamped firmly round its edge to ensure good thermal contact with the holder. Specimens are sometimes prepared from bulk material in the form of a disc of the same diameter as a support grid with a thin area at the centre. The thin part of the specimen is supported by the thicker part around the edge and the disc is clamped directly into the holder. The holder locates in a stage in the microscope which can be moved about, enabling different areas of the specimen to be studied.

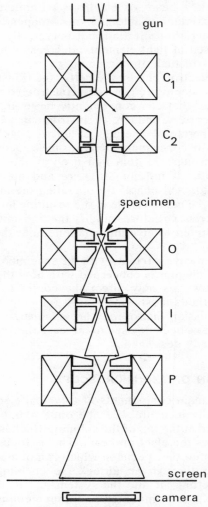

Figure 4.1. A schematic representation of the layout of the main components of a typical electron microscope. The electron gun is at the top of the column and C_1 and C_2 are the two condenser lenses. The second condenser lens has a variable aperture. The specimen sits in a holder that is inserted into the microscope through the airlock which lies between the second condenser lens and the objective lens O. This lens also has a variable aperture and there is another variable aperture in the object plane of the intermediate lens I. The projector lens P forms the final image on the fluorescent screen, which is observed through a glass window in the column. A plate or film camera below the screen enables the image to be recorded

Lenses below the specimen magnify the image formed by the electrons transmitted through the specimen, the final image being focused onto a fluorescent screen at the bottom of the column. Plateholders or film packs are stored in a magazine below this screen to enable the image to be recorded photographically after the screen has been raised out of the way. Certain of the lenses are provided with variable apertures, (usually in the form of a slide containing several fixed apertures, any one of which can be brought into position), and there are several sets of deflecting coils to enable the instrument to be aligned. Controls that pass through vacuum seals enable routine operations such as changing apertures or taking photographs to be carried out while the instrument is running.

4.1.1 The Illumination System

The design of the electron gun is similar to that of the scanning microscope described in chapter 2, the minimum cross-section of the beam produced at its crossover point being about 100 μm. For the reasons described below two lenses are normally used to focus the electrons at the crossover point onto the specimen.

In the first place, it is desirable to illuminate only that part of the specimen that is under observation. This is because the beam can adversely affect the specimen by heating it, by producing radiation damage or by causing a layer of contamination to form on the surface of the specimen as a result of residual organic vapours in the column being decomposed by the electron beam. Reducing the illuminated region reduces the area over which such damage occurs. However, in order to reduce the size of the beam sufficiently for normal use it is necessary for the condenser lens system to have an overall demagnification of about 50. If a single condenser lens were used to produce this reduction in size, the distance from the lens to the specimen would have to be very small, making it difficult to accommodate the specimen-changing airlock mechanism. In the two-lens arrangement, the strong first lens produces a strongly demagnified image of the electron gun crossover which is then projected with little change in size onto the specimen by the weaker second lens. This second lens is supplied with a variable aperture which controls the intensity of the illumination at the specimen.

Because the second condenser lens has a long focal length it has a large amount of spherical aberration, and the area of specimen illuminated can never be reduced to the size expected from geometrical considerations, being limited to a minimum diameter of about 1 μm. This lens also suffers from astigmatism due to the effective field not being circularly symmetrical, so that it acts like a combination of a normal positive lens and a weak cylindrical lens (see section 4.2.4). As a result, the illuminated area of the specimen is increased in size at exact focus, and on one side of focus is elongated in a direction parallel to the axis of the cylindrical component of the lens, being elongated in the direction perpendicular to this on the other side of focus. Astigmatism is corrected by passing currents through small coils arranged round the microscope axis so that the field they produce cancels the non-cylindrical component previously present.

As indicated in figure 4.1 a large fraction of the electrons brought to a focus by the first condenser does not pass through the second condenser lens because the beam divergence at this point is so high. Moving the image due to the first lens closer to the second does increase the number of electrons reaching the specimen, but as this also increases the magnification produced by the second lens these electrons are distributed over a larger area and there is no gain in brightness. Increasing the size of the aperture in the second condenser lens obviously permits more electrons to pass through it, but in this case, because of the aberrations of the lens the size of the illuminated area at focus also increases. Consequently, while there is an increase in image intensity when working at low magnifications, at high magnifications, when the area of the specimen being studied is small, the minimum illuminated region is much larger than the area under observation, and most of the extra electrons are not available in forming the image. On the other hand, the heating effect of the beam is now much greater, so that it is generally undesirable to work with a larger aperture than necessary.

The only satisfactory way of increasing the image intensity is to improve the brightness of the gun. Some improvement occurs when using a pointed filament, while a substantial increase is given by a field-emission gun, which however has the major drawback of requiring a very high vacuum.

4.1.2 The Imaging System

The imaging system typically consists of three lenses. The first (objective) lens has a short focal length and is provided with an aperture in the back focal plane of the lens (at A in figure 4.2), where parallel rays reaching the lens are brought to a focus. Another aperture is located in the normal image plane of the lens (at B). Two other lenses, the intermediate and projector lenses, successively magnify the image further to give a final magnification which may typically lie in the range 1000–100 000. Since the position of the specimen is fixed, it is not possible appreciably to change the magnification produced by the objective lens. If the current through the objective lens is altered slightly, the position of the image it produces and its magnification will also alter slightly, as shown in figure 4.3. However, as far as the intermediate lens is concerned, this constitutes a much larger relative movement of its object plane, with, as shown in the figure, a consequent large change, on refocusing this lens, of its magnification. Most of the variation in magnification of the microscope is obtained in this manner, although in practice the adjustments are performed in the reverse order. The intermediate lens current is altered until the object, now a little out of focus, is seen at about the magnification required, when the image is brought back into focus by adjusting the objective lens setting.

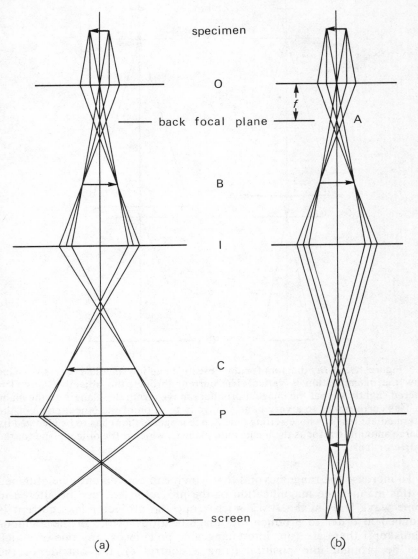

specimen

O

back focal plane

A

B

I

C

P

screen

(a)   (b)

Figure 4.2. Ray paths for electrons (a) forming an image and (b) forming a diffraction pattern on the viewing screen. To obtain the diffraction pattern the intermediate lens current is reduced to make the plane A, where the diffraction pattern is formed, conjugate to plane C, which is where the intermediate lens normally forms an image of the specimen. The projector lens enlarges the image or diffraction pattern in this plane onto the viewing screen

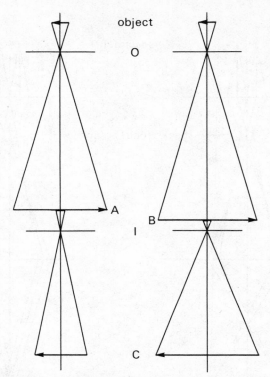

Figure 4.3. A ray diagram for the objective and intermediate lenses showing how the magnification is varied. The current through the objective lens O is altered slightly so that the image it produces moves from the plane A to the plane B. This corresponds to a very large change in position of the object plane of the intermediate lens I, whose setting (and hence magnification) has to be changed by a large amount to refocus the image onto plane C, which is the object plane for the projector lens

To increase the range beyond that which can conveniently be obtained in this manner the magnification of the projector lens can be altered in some way. Thus in the JEM 7 microscope the projector lens current is reduced in order to produce low magnifications, while in the Siemens Elmiskop 1 there are four interchangeable polepieces, any one of which can be brought into position using a control on the outside of the microscope, giving four different magnification ranges.

4.1.3 Diffraction Pattern Observation

If the microscope specimen is crystalline then all the suitably oriented sets of lattice planes will scatter electrons into corresponding diffracted beams. If the second condenser lens current is sufficiently reduced parallel illumination will fall on the specimen, so that the electrons in any given diffracted beam will be travelling parallel to each other. As they are brought to a focus in the back focal plane of the objective lens, a sharp diffraction pattern will be formed there. By reducing the intermediate lens current sufficiently this plane can be focused onto the projector lens object plane with the result that an enlarged image of the diffraction pattern appears on the viewing screen. A ray diagram corresponding to this situation is shown in figure 4.2b, and in figures 4.4a and b are shown an image and the corresponding diffraction pattern of an aluminium alloy specimen containing precipitates.

In the absence of any apertures in the system the whole of the illuminated area of the specimen contributes to this diffraction pattern. If, however, a magnified image of the specimen is produced on the viewing screen and the selected-area aperture (at B in figure 4.2) is inserted then, since this aperture is in the object plane of the intermediate lens, it also will be in focus. As a result, the viewing screen will appear dark except for the part corresponding to the magnified image of the aperture, within which part of the specimen image will be visible. This is seen in figure 4.4c which is a double exposure with and without the selected-area aperture. The circular area of the specimen visible with the aperture in place (part of one of the precipitates) can be seen against the darker image of the rest of the specimen which received electrons only during the exposure before the aperture was inserted. If the microscope is now adjusted to show the diffraction pattern then, since the objective lens setting is not changed, this pattern is formed only by those electrons passing through the aperture at plane B and thus corresponds to the selected area visible on the screen with the aperture in position. The diffraction pattern from the area selected in figure 4.4c is shown in figure 4.4d, where it can be seen that some of the spots previously present are now absent. That these diffracted beams come from the precipitate alone can be checked by using the objective lens aperture to select one of them to form an image. The one selected has been shown up by a double exposure as before and can be seen in figure 4.4d as the spot with a disc of greater background intensity around it. In the image formed with this beam, figure 4.4e, only the precipitate previously selected has any appreciable intensity, con-firming that this diffracted beam comes from the precipitate and not the matrix crystal. The diffraction pattern from the material to one side of the precipitate is at figure 4.4f, and it can be seen to be entirely different from 4.4d.

Thus the diffraction facility makes it possible to observe the diffraction pattern from a small selected area of a specimen, and so determine its orientation, and also to locate the objective aperture accurately by observing it and the diffraction pattern as it is positioned.

In practice, when the selected-area aperture has been inserted the intermediate lens is adjusted to bring it into focus, after which the objective lens setting is altered to bring the image back into focus. This ensures that the objective lens image and the selected-area aperture lie in the same plane (they are in focus simultaneously) so that the area selected is sharply delineated. This condition is obtained at only one particular

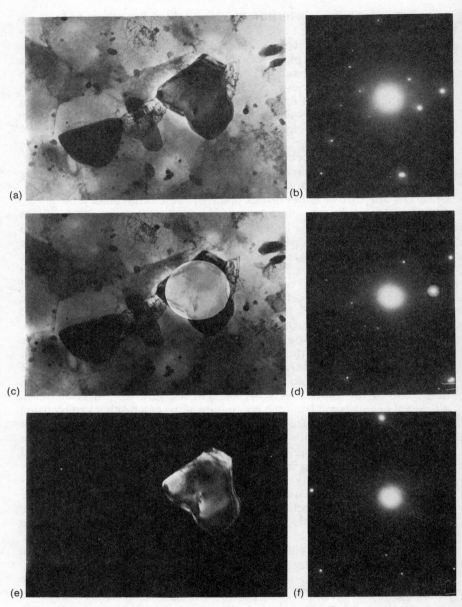

Figure 4.4. Micrographs and diffraction patterns of an over-aged Al–Cu 4 per cent alloy: (a) and (b) are a micrograph and the corresponding diffraction pattern of an area containing several precipitates; (c) is a double exposure, with and without the selected-area aperture, showing the area of precipitate used for the selected-area diffraction pattern; (d) diffraction pattern corresponding to the aperture position in (c); this is also a double exposure, with and without the

magnification, although of course this has no effect upon the size of the diffraction pattern which is subsequently formed.

4.2 RESOLUTION OF THE ELECTRON MICROSCOPE

4.2.1 The Theoretical Limit

The ultimate resolution attainable with an optical microscope, since lenses can be made effectively free of aberrations, is normally limited by the angular aperture of the objective lens (semi-angle α), and the wavelength λ of the light employed. On the Rayleigh criterion two self-luminous points can just be resolved at a separation δ given by

$$\delta = 0.61\lambda/\sin \alpha \qquad (4.1)$$

Thus δ is typically of the order of λ, the wavelength of the light used. This equation shows that if lenses of large aperture could be designed for the electron microscope a resolution of about 5 pm would be achieved at an accelerating voltage of 100 kV. In practice, the resolution is at best some two orders of magnitude worse than this, due mainly to the fact that magnetic lenses inherently suffer from severe aberrations which, owing to the impossibility of producing divergent lenses, cannot be corrected. Other factors which contribute to the overall lack of resolution include instabilities in the lens currents and the accelerating voltage, mechanical vibration of the column and drift of the specimen stage. The instrument is designed to ensure that these factors have a smaller effect on the resolution than the lens aberrations. As far as the overall resolution of the microscope is concerned it is the aberrations of the objective lens that are the most important, the influence of any other lens in the system being reduced in proportion to the magnification of the image which it uses as its object.

The effects of the main aberrations of the objective lens are now considered.

4.2.2 Chromatic Aberration

This is the dependence of the focal length of a lens upon the wavelength of the electrons passing through it, and arises because the faster an electron travels the less it is deflected by any particular magnetic field. If the accelerating voltage V of the microscope changes by ΔV then the corresponding change Δf in a lens of focal length f is given by

$$\Delta f/f = K\,\Delta V/V \qquad (4.2)$$

---

objective aperture, showing the diffracted beam selected for a dark-field micrograph; (e) the dark-field micrograph given by the beam selected in (d); only the precipitate has appreciable intensity, confirming that the diffracted spot comes from the precipitate and not the matrix; (f) diffraction pattern of adjacent matrix crystal

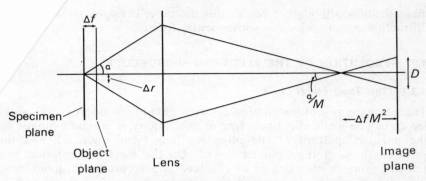

Figure 4.5. Chromatic aberration changes the effective focal length of a lens by Δf for electrons with a given wavelength change. This is roughly equivalent to a change of $-\Delta f$ in specimen position instead, and hence it can be seen that these electrons are focused at a distance of approximately $M^2\Delta f$ from the normal image plane. In the image plane they give rise to a disc of diameter D, where $D = 2\alpha M\Delta f$, corresponding to a disc of radius $\Delta r = \alpha\Delta f$ in the object plane

K is a constant close to unity. From figure 4.5 it can be seen that the size of the disc in the image produced by this change Δf is equivalent to a disc of radius Δr at the object plane given by

$$\Delta r = \alpha\,\Delta f \qquad (4.3)$$

In order that chromatic aberration should not impair the resolution it is necessary that this disc be smaller than the resolving power of the microscope. Taking $K = 1$, $\alpha = 10^{-2}$ radians and $\Delta r = 200$ pm (equivalent to a best resolution of about 400 pm), this means that the fractional range of electron energies allowable, $\Delta V/V$, must be less than about 10^{-5}; that is, the electron beam must be highly monochromatic with an energy spread of no more than 1 eV for an accelerating voltage of 100 kV. (A similar solution to the problem of chromatic aberration is sometimes found in optical microscopy, where filters giving monochromatic light are used for high-resolution work).

However, when the electrons pass through the specimen some of them are scattered inelastically and lose energy (typically 10 eV or more), so that they are not brought to a focus in the same plane as those that have lost no energy. This limits the thickness of material that may be studied at high resolution to about 20 to 50 nm. Thicker specimens can be studied at lower resolutions, until a point is reached where most of the intensity is lost from the image into the diffuse background.

4.2.3 Spherical Aberration

The wavefront leaving a perfect lens would be spherical in shape, centred upon the image point onto which it converges, as shown in figure 4.6. The effect of spherical aberration is to cause the lens to refract the rays

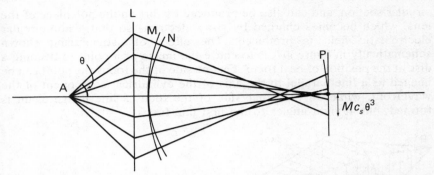

Figure 4.6. Spherical aberration causes rays passing through the lens at an angle θ to be bent more strongly than by a perfect lens. The real wavefront N is distorted compared with the perfect wavefront M which is a part of a sphere centred and converging upon the image point P. The separation of the two wavefronts is $\frac{1}{4}c_s\theta^4$, so that a ray leaving A at θ crosses the image plane at a distance of $Mc_s\theta^3$ from P, the point in the image corresponding to A. The disc thus formed in the image is equivalent to a disc of radius $c_s\theta^3$ in the object plane

passing through the edges of the lens more strongly than a perfect lens, so that the actual wavefront has the form indicated. The separation d of these two wavefronts at the lens is given as a function of θ by

$$d = \tfrac{1}{4}c_s\theta^4 \tag{4.4}$$

where c_s is the spherical aberration coefficient for the lens.

As a consequence of this greater refraction, the ray leaving the object at angle θ to the axis is bent so as to cross the image plane at a distance from the ideal image point equal to $c_s\theta^3$ in object coordinates. A point in the object is thus blurred into a disc in the image, the size of this disc being proportional to the cube of the aperture of the lens. Reducing the aperture size increases the definition until the spreading of the image point by diffraction at the aperture becomes important. These effects are comparable for an aperture of semi-angle α where

$$c_s\alpha^3 = 0.61\lambda/\alpha \tag{4.5}$$

For $c_s = 2$ mm and $\lambda = 3.7$ pm this gives a value for α of about 6×10^{-3} radians, and hence a best resolution of approximately 400 pm. As the spherical aberration coefficient c_s is comparable in magnitude to the focal length, the objective lens is constructed with the shortest possible focal length, even though this restricts the space available for the objective aperture and the specimen.

4.2.4 Axial Astigmatism

Axial astigmatism corresponds to a cylindrical component being imposed upon the spherical wavefront described in the previous section. It arises because the bore of the objective lens cannot be machined to a truly circular section, and can also be produced by dirt on the polepiece of the lens, which becomes charged by stray electrons so that a non-circular electrostatic field is produced. The effect of astigmatism, shown schematically in figure 4.7, is to cause the image of a point to become a disc at the position of best focus. On one side of the focus the point can be imaged as a line parallel to the axis of the cylindrical component of the wavefront, while on the other side of the focus a line image again is formed, but at right angles to the first line focus.

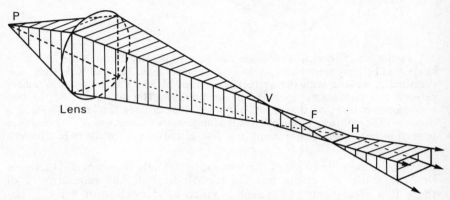

Figure 4.7. An astigmatic lens has a cylindrical component which causes the rays leaving an object point P to be focused as a vertical line at V on one side of focus and a horizontal line at H on the other side. The image of the point is a disc at the focus F, which is approximately halfway between V and H. Consequently such a lens will produce a sharp image of a vertical line at V, whereas a horizontal line will appear sharp at H

Axial astigmatism can be corrected by cancelling the non-circular part of the lens field, usually with a set of coils arranged around the axis of the instrument, the currents being adjusted to produce the necessary field direction and strength. Correction is usually carried out with the aid of a thin carbon film with holes in it (a 'holey film') which is observed close to exact focus at high magnification. Under these conditions one or more fringes, known as Fresnel fringes, are observed around the edge of each hole. Their formation can be understood with the aid of figure 4.8, which shows the situation where parallel illumination of wavelength λ falls on the edge of an opaque screen and casts a shadow on the plane P located at a distance d beneath. Fringes parallel to the edge of the screen are produced in the region of the edge of the geometrical shadow by interference between the waves which pass close to the screen (see Longhurst, 1957, p. 272). The positions of these fringes are given approximately by

$$X_n = (2nd\lambda)^{\frac{1}{2}} \tag{4.6}$$

where $n = 1, 2, 3$, etc.

In the electron microscope the material around the edge of a hole in the

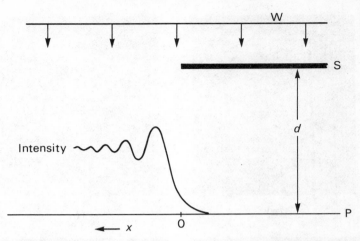

Figure 4.8. When a plane wave W falls on an opaque screen S, interference between the waves passing the screen causes fringes to be observed in a plane such as that at P, near to O, the position of the geometrical shadow of the screen

specimen acts as the screen, but because it is not completely opaque the form of the fringes is not quite the same as predicted by the simple theory (for example, Hall, 1953, chapter 10). By focusing the instrument on a plane distance d below the specimen, fringes similar to those described above may be observed, the first fringe appearing at a distance of about $(2\lambda d)^{\frac{1}{2}}$ from the edge. In practice, due to lack of coherence in the electron beam, only this first fringe is usually visible, as is the case in figure 4.9a. At exact focus no fringes are seen, and the edge of the specimen appears poorly defined (figure 4.9b), while on the other side of focus a fringe with reversed contrast is produced (figure 4.9c). When the lens is astigmatic the resolution and form of these fringes will vary around the edge of the hole. As can be seen from figure 4.7, a line parallel to the axis of the cylindrical component of the lens can be brought to a focus at one lens setting away from exact focus, while lines perpendicular to this are sharp at a setting on the other side of focus. Thus near focus one part of the edge of a hole will be underfocused and show one type of fringe while at 90° away around the hole the edge will be overfocused and show a fringe with reversed contrast, as in figure 4.9d. To correct the lens, the stigmator controls are adjusted until the most uniform fringe contrast around the edge of a hole is achieved.

4.2.5 Depth of Field and Depth of Focus

The depth of field of the microscope is the distance along the axis through which the object can be moved without noticeably reducing the resolution. If the specimen is moved a distance ΔL away from the plane on which the objective lens is focused then a point in the object will be imaged on the viewing screen as a disc. From figure 4.10 it can be seen

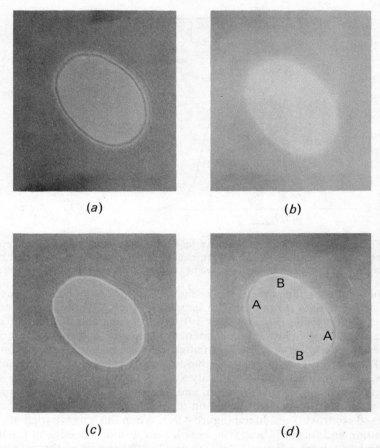

Figure 4.9. Fresnel fringes around a hole in a carbon film: (a) overfocus; (b) close to focus; (c) underfocus; (d) near focus, the form of the fringe showing that the objective lens has considerable astigmatism

that a disc of this size would be given by a disc of diameter $2\Delta L\alpha$ in the object plane. Thus the displacement ΔL produces a blurring of detail in the image equivalent to a blurring of $2\Delta L\alpha$ in the specimen, and if this is not to reduce the microscope resolution it must be less than the finest detail resolvable. If α is assumed to be 6×10^{-3} radians and the resolution is taken as 400 pm then ΔL must not exceed about 40 nm for this condition to hold, giving a depth of field $2\Delta L$, through which the specimen may move, of approximately 80 nm. This is greater than the usual thickness of specimens used for very-high-resolution work. More typically a resolution of around 2.5 nm is acceptable, in which case the allowable depth of field is 400 nm. This again is greater than the thickness of specimens that can be used at this resolution, showing that under

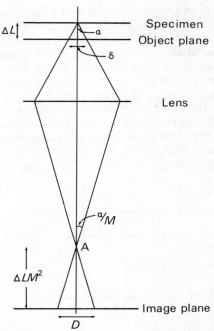

Figure 4.10. If the specimen is displaced a distance ΔL from the true object plane of the lens, then rays leaving a point on the specimen within a cone of semi-angle α pass through (or appear to come from) a disc of diameter $\delta(= 2\alpha\,\Delta L)$ in the object plane. This is imaged as a disc of diameter D in the image. ΔL must be small enough to keep δ less than the resolution required. Moving the object a distance ΔL corresponds to moving the image plane by $M^2\Delta L$

normal conditions the depth of field is sufficient to enable points at all depths within a specimen to be resolved simultaneously.

The depth of focus of the image is the distance through which the screen receiving the image can be moved without affecting the resolution, and can be calculated in a similar manner. Simple geometrical arguments show that if the linear magnification of a system is M, then when the object point moves by ΔZ the conjugate image point moves a distance $M^2\Delta Z$. Therefore, if the depth of field in the object plane is $2\Delta L$, the depth of focus of the image is $2M^2\Delta L$, which for $M = 2 \times 10^4$ and $\Delta L = 400$ nm is 160 m. In consequence, it is not necessary to refocus the microscope in order to record a sharp image on a photographic plate located a few centimetres below the level of the viewing screen.

4.2.6 Accuracy of the Selected-area Diffraction Technique

The minimum area from which a selected-area diffraction pattern can be obtained is determined by the spherical aberration of the objective lens.

This has the effect, as explained in section 4.2.3, of causing rays inclined at an angle to the optic axis to be deviated more strongly than they would be by a perfect lens. Thus, if the image of a crystalline specimen formed using the unscattered electron beam is brought to a focus in the plane of the selected-area aperture, then the image of the same area formed using a diffracted beam is brought to a focus in a plane some distance above this aperture. Because of this diperence in level and also the related inclination of the diffracted beam to the axis, those electrons which come from the same point in the specimen but in different diffracted beams, do not pass through the same point in the aperture plane, as shown in figure 4.11. The images are displaced in this plane from the zero-order image by a distance d_1 equal to $Mc_s\theta^3$ (where M is the magnification), corre-

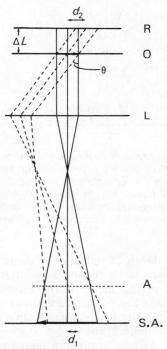

Figure 4.11. A schematic representation of the effects of defocus and spherical aberration on the accuracy of the selected-area diffraction technique. The image of the object at O is focused on the selected-area aperture plane using the undeflected electron beam. The rays diffracted at an angle θ to the axis are brought to a focus at A some distance above the first image, and they pass through the selected-area aperture plane at a distance d_1, from the corresponding undiffracted ray, where $d_1 = c_s M\theta^3$, equivalent to $c_s\theta^3$ in object coordinates. If, in addition, the object is moved a distance ΔL away from the object plane, those diffracted rays that pass through the centre of the object plane originate from an area that is displaced from the axis by a distance $d_2 = \theta\Delta L$. The total displacement due to both effects is, in object coordinates, $\theta\Delta L + c_s\theta^3$

sponding to an effective displacement d in the object plane of $c_s\theta^3$. For a typical value of $c_s = 2$ mm the images formed using the 200, 400, etc. set of reflections from copper at 100 keV are displaced by the following amounts

reflection	200	400	600	800
displacement in object coordinates	16 nm	0.12 μm	0.44 μm	1.02 μm

This means that if an aperture is used that selects an area of the specimen 1 μm in diameter, then the spots from the 200 and 400 beams will come almost entirely from the area selected. The 600 spot will come from an area about half of which lies outside the selected area, while the 800 spot will come from another area, adjacent to the one selected. This consequently sets a lower limit of about 1 μm on the size of area that can be studied with this technique.

This treatment assumes that the specimen is in focus for the zero-order beam, whereas in practice it is difficult to adjust the focus at the specimen level to better than a few μm, owing to a lack of fine detail observable in thicker specimens. This gives a further displacement to the area from which the diffraction spots come, as can be seen in figure 4.11, which shows that the additional displacement is $\theta\Delta L$. Again, considering copper at 100 keV, when the focusing error is 5 μm the value of $\theta\Delta L$ is 0.1 μm and 0.4 μm for the 200 and 800 reflections respectively. Thus the distance between the area giving rise to a diffraction spot and the area selected using the transmitted beam may be larger than the value given in the table.

Both of these problems can be overcome if the condenser lens system is modified by including extra lenses so that the illuminated area of the specimen can be reduced to a very small size (100 nm or less across; Chapman and Stobbs, 1969). In this case, since only the small illuminated area can give rise to the diffraction pattern, no selecting aperture with its consequent problems need be used, and a diffraction pattern from the small, illuminated area alone is obtained.

4.3 CALIBRATION OF THE MICROSCOPE

4.3.1 Magnification Determination

The magnification of an optical microscope depends upon the focal lengths of the lenses used and their separations, and may, in general, be found for any given combination of lenses by referring to a chart. The electron microscope has lenses whose focal lengths are variable, but in principle it ought to be possible to calibrate the total magnification as a function of the currents through the image-forming lenses. However, because of hysteresis effects in the soft iron polepieces and because the specimen level may vary significantly from one specimen holder to another, this gives only an approximate value for the magnification for any given set of conditions. While this may be adequate for most purposes, it is necessary for accurate quantitative work to calibrate the microscope for the experimental conditions of each working session. The principle of any magnification determination is to photograph the image of an object that contains detail of a known size: the ratio of the size of this detail on the photograph to that in the object is the magnification. The most widely used test objects are

(a)   For very low magnifications of up to a few hundred, a microscope specimen support grid, whose grid-bar spacing has been independently measured with an optical microscope, may be used.

(b)   A shadowed carbon replica of an optical diffraction grating may be made in the way described in section 4.4.1. This may typically have 2000 lines/mm and is a convenient and stable test object for magnifications in the range of 1000–10000.

(c)   Polystyrene latex spheres whose diameters vary only slightly about their average value are available as a suspension in water that can be coated onto a thin substrate and allowed to dry. Typical mean sizes lie in the range of 0.1–0.5 μm, so that latex spheres are useful for measuring magnifications in the range of 5000–50000. Some spheres on a carbon substrate are shown in figure 4.12. They have the advantage that they can be deposited upon the surface of the specimen itself, so that the magnification is determined at the same time as the features of interest in the object are recorded.

(d)   For very high magnifications test objects of atomic dimensions are needed, usually in the form of crystals with known lattice spacings. Molecular crystals with relatively large lattice spacings are often used, such as catalase, a beef liver enzyme which has a lattice spacing of 8.12 nm. Inorganic crystals with smaller spacings are also used, and tests using materials with interplanar spacings of several hundred pm or less serve both as magnification checks and resolution tests.

4.3.2   Rotation

Because the electrons travelling through a magnetic lens follow roughly helical paths, each lens produces a rotation of the image relative to the object, the amount of the rotation depending upon the currents passing through the lenses. Thus as the lens currents, and hence the magnification, are changed the image not only alters in size but also rotates. Since it is often necessary to know this rotation, and in particular to know the rotation of the final image relative to the corresponding diffraction pattern, a method of measuring it is required.

An approximate value of this relative rotation may be found by photographing the image for several settings of the intermediate lens as it is changed from the imaging to the diffraction condition. This gives a rough value of the relative rotation for any intermediate lens current and establishes that it is equal to a certain number of multiples of π plus a small part (less than π) which is known only approximately. It can be found accurately using a suitable specimen material such as MoO_3, which can be prepared as thin needle-shaped crystals with their long edges

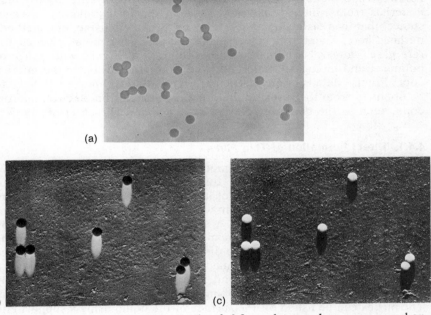

Figure 4.12. Electron micrograph of 0.3 μm latex spheres on a carbon substrate: (a) as-prepared, the spheres (which have a Fresnel fringe around them) look like discs, and none of the surface structure of the film is visible; (b) is a micrograph of a similar specimen that has also been shadowed: the enhanced contrast of the spheres and the presence of contrast from the surface of the substrate are now clearly visible. Printing (b) with reversed contrast, as at (c), makes the 'shadow' regions dark, and greatly enhances the three-dimensional effect

perpendicular to [100]. The crystal and its diffraction pattern are photographed on the same plate, and the small part is the angle between the edge of the image of the crystal and the appropriate direction in the diffraction pattern. The final value for the rotation is in fact incorrect by π, since the objective lens reverses the image relative to the object as well as rotating it, whereas it only rotates the diffraction pattern but does not reverse it. Subsequent lenses reverse both the image and the diffraction pattern in the same manner.

4.4 SPECIMEN PREPARATION METHODS

A very large number of methods, many differing only in detail, have been devised for preparing specimens suitable for electron microscopy. Most of these methods involve either preparing the specimen directly in the form of a thin film, or else reducing the thickness of a piece of bulk material until some regions are thin enough to transmit electrons. For completeness, some of the more important techniques are summarised below. While metals and organic tissues are probably the easiest bulk materials from which to make microscope specimens, almost any substance that is reasonably stable can be prepared using the methods outlined here. Materials as diverse as concrete, glass, metal, plastics, the rare gases, minerals and carbon fibres, as well as a very wide range of botanical and biological materials have been studied under the microscope. Precise details of any particular preparation technique will have to be obtained from the research literature, or manuals on electron microscopy, such as the books by Kay *et al.*, 1965, and Hirsch *et al.*, 1965.

4.4.1   Direct Formation of Thin Films

(i)   *Evaporation.* A small amount of the substance is heated in a vacuum of better than about 10^{-4} torr so that it evaporates. Methods of heating include making a wire of the material and passing a current through it, coating it onto a filament which is then heated in this manner or putting it into a heated crucible. A suitable substrate positioned nearby becomes coated with a layer of the material, the final thickness depending upon the rate of evaporation, the distance from the source and the time for which evaporation is continued. The film is cut into pieces of a convenient size by scoring it, and freed from the substrate either by floating it off in water or by dissolving the substrate in a suitable solvent. The pieces can be fished off the surface of the liquid using a support grid which is then clamped in the microscope specimen holder. Some materials such as carbon give amorphous films, but in most cases a polycrystalline specimen results. If the substrate material is suitable then the layer will deposit epitaxially (for example, gold on cleaved rock salt) giving a highly imperfect single-crystal film.

(ii)   *Surface replicas.* The surface morphology of a specimen may be studied at high resolution by making a replica of the surface which is thin enough for microscopy. Replicas can be made by processes involving one or two stages.

In the single-stage method a thin coating is applied to the surface as indicated at (i) and (ii) in figure 4.13a. For routine work a drop or two of a solution of a plastic such as 'Formvar' is spread over the surface and allowed to dry, while for high-resolution work an evaporated carbon film is frequently used. This coating, which is a replica of the surface, is stripped from the surface as at (iii) and cut into pieces for examination in the microscope.

If the replica cannot be removed from the surface without damaging it a two-stage process is used. The surface is coated with a rather thicker plastic film, which because of its greater strength can be stripped from the surface by pulling it off with adhesive tape or by floating it off in water. This cast of the surface is now coated with a layer of carbon as shown at (iv) in figure 4.13b, and the plastic dissolved away to leave a carbon replica, as at (v).

The contrast produced by a replica is not usually very high, but it may be increased by coating either the completed replica, or else the surface

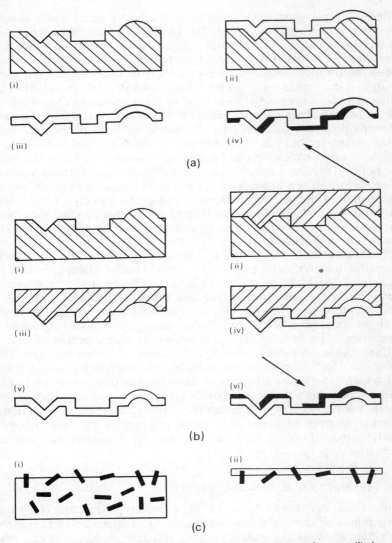

(a)

(b)

(c)

Figure 4.13. (a) Single-stage replication process. The surface, at (i), is coated with the replicating material, (ii), which is stripped from the surface to give a replica (iii). 'Shadowing': that is evaporating a metal of high atomic number onto the surface at a low angle causes surface irregularities to cast shadows and so become visible. (b) In the two-stage process the first replica, at (iii), is much thicker (and hence stronger) than in the single-stage method. A thin replica suitable for transmission microscopy is made of this replica, as at (v), and shadowed if required. (c) An extraction replica requires a surface through which second-phase particles project, as at (i). The replicating material is coated onto the surface and the solid dissolved, leaving some of the particles embedded in the replica as at (ii)

itself before replication, with a thin layer of a heavy metal which absorbs electrons more strongly than does the replica material. It is evaporated onto the surface at an oblique angle, and as can be seen from figure 4.13, this produces a deposit of the heavy metal which shows up the surface relief in the form of 'shadows'. It is evident from figures 4.13a and b that the single-stage replica has a surface that is complementary to that of the original sample, whereas the two-stage process results in the initial and replica surfaces having similar relief. If the surfaces are fairly smooth, shadowing produces contrast which is much the same in the two cases, whereas rough surfaces can give rise to very different final images. As an example of how shadowing can produce additional contrast in the image some polystyrene latex spheres on a carbon substrate, without and with shadowing, are shown in figures 4.12a and b. The greater contrast, and the enhancement of the three-dimensional form of the object, is immediately apparent in figure 4.12b. From the lengths of the shadows the shadowing angle (which may vary from one point to another in a large or buckled replica) can be calculated.

A further development of the technique may be used to study inclusions within a material, by preparing a surface (as by etching) so that the inclusions project above the surface, as in figure 4.13c. The replica material is coated onto the surface and the solid material is dissolved away, leaving some of the inclusions embedded in the replica. Their size and shape may be studied, and if thin enough can also be viewed in transmission. This type of replica is known as an extraction replica.

(iii) *Other techniques.* Further methods of preparing thin films suitable for electron microscopy include: electrodeposition onto a substrate which can then be dissolved to leave the film (for example iron onto copper); producing a thin surface oxide layer and then dissolving the solid beneath (used especially with aluminium and its alloys); or the reaction of a suitable mixture of gases to make a product in the form of small crystals, some of which can be collected on a microscope grid for examination.

4.4.2 Preparation from Bulk Samples

(i) *Cleaving.* Many crystals having a layer-type structure (for example mica) may be cleaved down to thicknesses of 20 nm under favourable circumstances. By sticking pieces of adhesive tape to each face of a sheet of the material and pulling them apart the sheet will usually cleave into two thinner sheets. The process is repeated until a piece is obtained which is thin enough.

(ii) *Crushing.* Brittle materials, like glass, which do not cleave may be crushed to give small particles, some of which are usually in the form of flakes thin enough for transmission microscopy.

(iii) *Microtoming.* The bulk specimen is supported by embedding it in a suitable material, and is then clamped in a holder. A sharp knife, usually made from diamond or from the edge of a carefully broken piece of glass, is passed across the face of the specimen which is moved forward by a controlled amount after each pass, so producing slices equal in thickness to the distance moved by the holder after each cut. This technique is widely used for biological materials that are easily cut without causing appreciable damage. Crystalline materials tend to shatter if they are brittle, while if they are softer severe plastic deformation occurs and large numbers of defects are produced.

(iv) *Chemical thinning*. Most substances can be dissolved in appropriate chemical reagents, which can therefore be used to reduce the thickness of a slice of material obtained from a bulk sample. Depending upon the nature of the material such a slice can be obtained by cleaving, sawing with a fine-toothed saw or an acid stringsaw, or cutting with a spark eroder. It is necessary that the material should be removed from the surface in a manner that eliminates initial roughnesses so that the thin areas that are finally produced are both adequate in extent and also free from local thickness fluctuations that can give spurious contrast. This may be difficult to achieve with two-phase materials, where the rates of dissolution of the two components tend to be different. The initial slice is often made in the form of a disc which fits the microscope specimen holder, and the reagent is directed in a jet onto the disc, producing a hole at the centre around which are areas thin enough for microscopy. The thick edge of the disc allows the specimen to be handled conveniently and clamped in the microscope specimen holder without damaging the thin regions.

(v) *Electro-polishing*. This is a widely used variant of the chemical thinning process, in which the specimen is made the anode in an electrolytic cell so that material is removed from it when a current passes. An additional problem with this method is that the rate of removal of material is most rapid where the density of electric field lines at the specimen surface is greatest. Protecting the edge of the specimen with insulating lacquer and correct electrode design can help to reduce this problem. Alternatively, a jet method may again be used in which the electrolyte is directed onto the sample from a metal orifice which acts as the cathode. As soon as a hole is made in the material preferential attack due to concentration of the field lines occurs at the edges, rapidly removing the thin areas around the hole unless the current is turned off. As in the case of chemical thinning, problems often occur with two-phase materials.

(vi) *Ion-beam thinning*. This method involves knocking atoms off the surface of the specimen by bombarding it *in vacuo* with high-energy ions (typically argon ions at about 5–10 keV), a process known as sputtering. The ion beam is focused to limit the area of attack, and while the rate of removal varies from one substance to another, the technique generally works well with two-phase materials. This is because the beam is directed onto the surface at a low angle, so that when the more easily removed areas become lower than the rest they are protected until the harder parts projecting from the surface are also eroded. The major disadvantage of the method is the high cost of the equipment necessary and the long time (typically up to 24 hours) needed to produce one specimen.

4.5 SPECIAL STAGES FOR THE ELECTRON MICROSCOPE

Any microscope specimen stage is required to hold the specimen at the correct level in the objective lens polepiece without vibration or drift in position, and at the same time permit the specimen to be moved around smoothly to enable different areas to be studied, while also allowing specimens to be changed readily through an airlock. In addition, in order to minimise the build-up of contamination on the sample, many stages are equipped with cold traps on which residual organic vapour condenses. When studying crystalline materials a further requirement arises due to the fact that most of the contrast in the image depends critically upon the diffracting conditions, and only by observing a specimen under several different sets of diffracting conditions can the maximum amount of information about the microstructure be extracted. Consequently, it is necessary to be able to tilt the specimen and so change the diffracting conditions, and all commercial instruments offer a tilting specimen holder as a standard item of equipment.

The desired specimen orientation is usually produced either about by tilting two mutually perpendicular horizontal axes, or about a single axis, the direction of which can be varied relative to the specimen—either by rotating the axis direction or the specimen. Tilts of up to $\pm 30°$ are usual, and with holders that have been made for special purposes much greater ranges have been obtained.

The electron microscope is frequently used to study the changes that have occurred as a result of some particular treatment, such as heating, straining or irradiation being given to a material. It may be helpful in understanding the details of the processes involved if the treatment, or something similar to it, can be carried out on a specimen that is simultaneously being observed in the electron microscope. In addition many materials, such as solidified gases, high-temperature phases of crystals, etc., are stable only under conditions considerably different from the normal environment in the microscope specimen holder.

Consequently, much effort has gone into designing specimen stages that, in addition to satisfying the criteria described previously, enable the operator to perform additional experiments in the electron microscope. A number are available commercially as accessories from microscope manufacturers. To give an indication of the potential versatility of the instrument some of these are outlined.

Heating stages enclose the specimen in a very small furnace, often enabling the specimen to be tilted as well, and have been used for studies of processes such as diffusion, annealing, phase transformations and decomposition. A variety of cooling stages, producing temperatures down to 4 K, have been designed and used for looking at low-temperature phase transformations, structures of condensed gases and their defects, etc. (see, for example, Venables and Ball, 1971).

By substantially modifying the airlock section and pumping it independently of the rest of the microscope, (for example, Valdrè *et al.*, 1970) it has been possible to obtain a very good vacuum at the specimen position. By evaporating materials directly onto a thin substrate in the specimen holder, condensation and the early stages of the formation of thin crystalline films have been studied.

Stages have also been made in which the specimen is clamped between grips so that it can be plastically deformed while the dislocation movement is recorded. Other stages have been designed for magnetic studies, in which the specimen is raised out of the magnetic field of the objective lens polepiece and held between sets of Helmholtz coils, so that the magnetic domain structure can be observed while the field applied to the specimen is changed. Still other stages have been designed to allow the X-rays emitted from the specimen to be counted, providing some of the facilities for X-ray microanalysis (for example, Chapman, 1968), or which permit the specimen to be observed while it is irradiated with ions or other fast particles, so that the resulting radiation damage is seen as it is produced.

Many other special stages have been constructed, and the potential range of applications is limited only by the restricted space available near the objective lens and by the ingenuity of the experimenter. It is worth noting that the larger dimensions of the lenses in the very-high-voltage microscopes now being produced provide more room for such special stages, and the thicker specimens which can be studied in these instruments are more representative of bulk materials. The availability of these microscopes thus greatly increases the potential scope of such *in situ* experiments.

4.6 TRENDS IN MICROSCOPE DESIGN

A number of difficulties that arise when using the electron microscope have been noted in passing, and the design of the instrument is continuously evolving to try to overcome them. The main problems are those of resolution, image brightness, degradation of the specimen by the beam and the inability to study thick objects. There is also the desire, noted in the previous section, to be able to use the microscope as a self-contained miniature laboratory.

The resolving power of the microscope is improved if c_s, the spherical aberration coefficient of the objective lens, is reduced. Since, for magnetic lenses, this quantity is roughly equal to the focal length there is clearly a limit to the reduction that can be expected, given that there has to be space available within the lens for the specimen and its holder. Moreover, since the resolution is proportional to $c_s^{-\frac{1}{4}}$, halving c_s will improve the resolution by only about 20 per cent, and any improvement due to reducing c_s is not likely to be very great. Objective lens focal lengths and

c_s values have both tended to become smaller in successive instruments, and this tendency will no doubt continue. Some manufacturers now offer their instruments in two forms: a high-resolution model with small c_s but little room in the polepiece for other than a standard holder, and a standard model having lower resolution but with space for tilting holders and other special specimen stages.

A high-resolution micrograph contains extra information in the form of interference effects that cannot be interpreted just by looking at it. It can, however, be recovered by processes outlined in section 5.3, so increasing the effective resolution. For this extra information to be available it is necessary that the lens currents and accelerating potential should be extremely stable, and recent high-resolution microscopes have greatly improved power supplies compared with older instruments. As part of the search for stability, attempts are being made to use superconducting coils for lenses.

To obtain high resolution, it is necessary to work at very high magnifications, in order that the fine detail in the object is large enough to be resolved on the viewing screen and photographic plate. Consequently the maximum magnification available has tended to increase in step with the resolving power, and some instruments (for example, the Philips EM 300) now have two intermediate lenses in order to be able to achieve the required magnification conveniently. Because of the problems described in section 4.1.1 it is difficult to obtain a sufficiently high beam intensity at the specimen to enable it to be observed with adequate brightness at such high magnifications. The situation is eased to some extent in high-resolution microscopes since the specimen is placed well inside the magnetic field of the objective lens, and the residual field above the specimen acts as an extra condenser lens. Some instruments include a small, third lens in the airlock section, which reduces the area of the focused electron beam, and manufacturers also produce pointed filaments for the electron gun which increase both the brightness and coherence of the illuminating beam. An alternative method of attack is to amplify an image that is too weak to be seen easily, and image-intensifying systems that are fitted in place of the plate camera are becoming available, the intensified image being displayed on a television monitor screen. This arrangement also enables materials that are easily damaged to be observed while using a low beam current.

Thicker specimens cannot be studied at high resolution because chromatic aberration prevents those electrons that have been inelastically scattered from being focused properly. One answer to this problem is to increase the accelerating voltage, in which case the inelastic scattering is reduced, and in addition any given energy loss constitutes a smaller fractional change in energy and hence in the wavelength of the electron. High-voltage microscopes show appreciable increases in penetration, but the potential increase in resolution has been less easy to realise on account of the fact that the higher the energy of an electron the less it is deflected by a magnetic field. The focal lengths of the lenses of high-voltage microscopes tend as a result to be longer than those of conventional instruments, and the aberration coefficients consequently larger. At the time of writing, the 3 MeV instrument at Toulouse is the most powerful microscope in operation.

Other approaches to the problem arising from energy loss and chromatic aberration involve building essentially new instruments. They include the energy-selecting microscope and the scanning-transmission electron microscope (STEM), and are described in chapter 8. The first of these uses a dispersive element to remove the inelastically scattered electrons, while the second scans the beam across the specimen like the scanning electron microscope, but collects the transmitted, rather than the reflected, electrons. Since there are no image-forming lenses, chromatic aberration is not important, and the size of the electron probe limits the resolution. By using, in effect, an objective lens as the condenser lens, and a field-emission gun to obtain high source brightness, resolution comparable with a conventional microscope is obtained for thin objects, while it is much better for thick specimens. However, although the instrument looks more like the SEM of chapter 2, contrast from crystalline materials arises in essentially the same way as in the conventional microscope. Contrast production is the subject matter of chapter 5.

4.7 GUIDE TO FURTHER READING

There are a number of books which treat the design and instrumentation of the electron microscope in more detail than has been possible here. Two more advanced books on electron lenses as such are by Grivet (1972) and Klemperer and Barnett (1971).

Books which deal more specifically with the design and operation of electron microscopes, but which also have sections on electron lenses, include those by Hawkes (1972) and Kay (1965).

Recent developments are described in articles which appear from time to time in various review journals: among these are *Practical Methods in Electron Microscopy* from 1972 onwards; *Advances in Optical and Electron Microscopy* 1966 onwards; and *Advances in Electronics and Electron Physics*.

REFERENCES

Chapman, P. F., (1968), *Proceedings of the Fourth European Conference on Electron Microscopy*, Rome; *Tip. Pol. Vat.*, 253

Chapman, P. F. and Stobbs, W. M., (1969), *Phil. Mag.*, **19**, 1015

Grivet, P., (1972), *Electron Optics*, Pergamon, Oxford

Hall, C. E., (1953), *Introduction to Electron Microscopy*, McGraw-Hill, New York

Hawkes, P. W., (1972), *Electron Optics and Electron Microscopy*, Taylor & Francis, London

Hirsch, P. B., Howie, A., Nicholson, R. B., Pashley, D. W., and Whelan, M. J., (1965), *Electron Microscopy of Thin Crystals*, Butterworths, London

Kay, D. H., (ed.) (1965), *Techniques for Electron Microscopy*, Blackwell, Oxford

Klemperer, O. and Barnett, M. E., (1971), *Electron Optics*, Cambridge University Press

Longhurst, R. S., (1957), *Geometrical and Physical Optics*, Longmans, London

Valdrè, U., Robinson, E. A., Pashley, D. W., Stowell, M. J. and Law, T. J., (1970), *J. Phys. E.*, **3**, 501

Venables, J. A. and Ball, D. J., (1971), *Proc. R. Soc.*, **A322**, 331

5

Electron Microscope Image Contrast

Ideally it should be possible to interpret any given electron micrograph unambiguously in terms of the corresponding specimen microstructure. However, image contrast can be produced by several different mechanisms and may depend critically upon the operating conditions of the microscope, such as the precise orientation of the specimen or the exact amount of defocus of the objective lens. Consequently a complete interpretation may be difficult even though the problem is usually simplified in practice by the fact that normally only one contrast mechanism is important. Although methods exist which in principle enable information about the object to be deduced directly from the image (for example, Head, 1969; Misell and Childs, 1972, 1973; De Rosier and Klug, 1968), these are not in general use since they require a substantial amount of computing time and work well only under favourable conditions. More usually, on the basis of general considerations of contrast production, a guess is made at a possible structure for the object by looking at a number of micrographs. This structure is then confirmed by calculating the contrast it would give under a set of specified conditions and comparing this with the corresponding image. Sometimes such a procedure does not lead to an unambiguous conclusion, but if the way in which contrast arises in each case is known, likely alternative structures may be distinguished by devising a set of operating conditions that would cause microscopes to give different images, and then observing the specimen under these conditions.

This chapter describes the different ways in which contrast may be produced, and shows how the contrast given by different types of substructure may be calculated.

5.1 ELECTRON SCATTERING

In order to set about calculating image contrast it is necessary to know how the electron beam interacts with the specimen. As mentioned in chapter 2, fast electrons can be scattered either elastically, in which case they lose no energy during the scattering process, or inelastically, when the scattered electron has less energy than it had originally and the specimen, or one atom in it, is left in an excited state. These two types of process normally affect contrast in different ways, and will be considered separately.

5.1.1 Elastic Scattering

A fast electron is scattered by a solid because of the electrostatic forces between it and the nuclei and electrons of the atoms. Since the state of any given atom is not altered during the elastic scattering process it is not necessary to know the details of its electronic structure, but only the form of its electrostatic potential $V(r)$ seen by the fast electron. Two alternative approaches are used in treating the elastic scattering. One considers the potential of the crystal as the sum of the potentials of the different atoms, and finds the scattering due to this total potential. The other calculates the scattering due to a single atom, and obtains the scattering from the solid by summing the separate contributions from the atoms in an appropriate manner. This second approach is similar to that usually employed in the case of X-rays, and assigns to each type of atom a scattering factor. These two approaches ultimately lead to the same result.

The scattering factor is calculated using the fact that a small volume $d\tau$, in which the potential is $V(r)$, scatters an electron wave isotropically, with a scattered amplitude dA at distance R given by

$$dA = \frac{A_0}{R}\left(\frac{2\pi me}{h^2}\right)V(r)\,d\tau \qquad (5.1)$$

A_0 is the incident amplitude and m and e are the mass and charge of the electron. The atom is represented by an assembly of these elementary volumes. Taking the incident and scattered wave vectors to be k_0 and k, the contribution of the volume shown in figure 5.1a to the scattered wave is $dA\exp(2\pi iK.r)$, where K, shown in figure 5.1b, is $k - k_0$. The exponential factor in this expression takes account of the difference in phase between the contribution scattered by the volume at r and that scattered at the centre of the atom. Neglecting any reduction in amplitude of the incident wave as it passes through the atom, and any further scattering of the scattered waves as they leave the atom (the Born approximation), the total amplitude A scattered by the atom is given by

$$A = \frac{A_0}{R}\left(\frac{2\pi me}{h^2}\right)\int_{atom} V(r)\exp(2\pi iK.r)\,d\tau$$
$$= (A_0/R)f(K) \qquad (5.2)$$

$f(K)$ is the scattering factor, given by

$$f(K) = (2\pi me/h^2)\int_{atom} V(r)\exp(2\pi iK.r)\,d\tau \qquad (5.3)$$

Figure 5.2 shows $f(K)$ plotted against K for several elements, using values taken from the *International Tables for X-ray Crystallography*

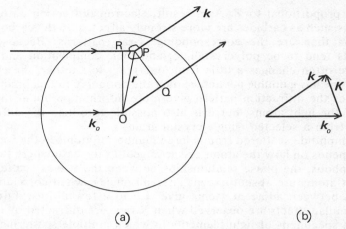

(a)   (b)

Figure 5.1. (a) The atom centred at O scatters electrons out of the plane wave of wave vector k_0. The contribution to the scattered wave with wave vector k from the small volume at r is not in phase with that scattered at the origin: the path difference is $OQ - RP$, corresponding to a phase difference of $2\pi(k \cdot r - k_0 \cdot r) = 2\pi K \cdot r$. (b) K is the change in wave vector of the electron when it is scattered, given by $k - k_0$

(1962). These scattering factors were computed using the rest mass m_0 of the electron in equation 5.3, rather than the correct relativistic mass. As a result they have to be multiplied before use by a correction factor, m/m_0; this is not negligible at typical microscope operating energies, being approximately 1.2 for 100 keV electrons.

The magnitude of the scattering factor increases slowly with atomic number, being roughly proportional to $Z^{1/3}$ for fast electrons: as is evident from figure 5.2 quite light atoms scatter electrons almost as well as do heavy atoms, unlike the case of X-rays where the scattering factor is

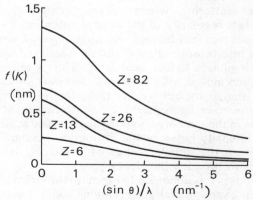

Figure 5.2. $f(K)$ as a function of $\sin \theta / \lambda$ for several elements ($K = 2 \sin \theta / \lambda$)

roughly proportional to Z. As a result, electron diffraction patterns of materials such as carbides are more strongly affected by the carbon atom positions than are the corresponding X-ray patterns. Because such materials tend to be polycrystalline, and their composition and hence structure often changes a little from one grain to another, the electron microscope has a double advantage over the normal X-ray technique. Not only does the diffraction pattern give more information about the positions of the light atoms, but it is also possible to obtain the diffraction pattern from a selected, single-crystal grain.

The amplitude scattered from a large number of atoms in the form of a solid depends on how the atoms are arranged. In the case where the solid is amorphous, the phase relationships between the waves scattered by different atoms are absent except at small angles, when correlations in position between adjacent atoms give rise to a few diffuse diffraction rings. Similar effects are observed when X-rays are diffracted by liquids. For thin specimens of light elements in which multiple scattering is not too important, the overall distribution of the scattered intensity, other than near to the diffuse rings, is similar to that from a single atom.

If the material is crystalline the atoms are regularly arranged, and constructive interference between the waves scattered elastically by different atoms gives rise to sharp diffraction maxima, the conditions for obtaining these being given, as in the case of X-rays, by Bragg's law. Unlike the X-ray case, however, the scattering is relatively strong and the effect of multiple scattering of the electrons has to be taken into account. A treatment of electron diffraction which does this is given in section 5.3.

5.1.2 Inelastic Scattering

The inelastic scattering from an isolated atom is less easy to calculate, since a detailed knowledge of the atomic electron wave functions is required. Furthermore, additional scattering processes become possible when atoms are aggregated into a solid. These include plasmon scattering, in which the conduction electrons are set into collective oscillations, and thermal diffuse scattering, which occurs because thermal vibrations destroy the perfect regularity of the crystal potential. To a first approximation these processes may be regarded as incoherent, so that they do not give rise to any interference effects, and the scattered electrons have no memory of their phases before scattering. This is not entirely correct, however, and such inelastically scattered electrons are capable of giving image contrast similar to, but weaker than, that given by the elastically scattered electrons (see figures 8.9d, e). Scattering involving the excitation of electrons in the ion cores, and thermal diffuse scattering, both give rise to diffuse intensity between the diffraction maxima. In the case of plasmon scattering the angular range of scattering is small, and this process appears to broaden the Bragg peaks.

For most purposes it is adequate to estimate the strength of the inelastic scattering from experimental observations. The electronic processes increase in strength roughly in proportion to Z (that is, in proportion to the number of electrons in each atom), while thermal scattering accounts for about one-third of the total effective absorption in light elements and for one-half or more in the case of heavy elements such as lead or gold.

5.1.3 Kikuchi Patterns

For most purposes the diffuse scattering due to the various processes described above is just a nuisance, tending to reduce the image contrast and definition, and limiting the thickness of material that can be studied. However, there is one effect arising from the diffuse scattering that is extremely useful when studying defects in crystalline specimens.

If the diffraction pattern of a crystalline specimen is observed in the electron microscope the diffusely scattered intensity can be seen in between the diffraction spots. This intensity would be expected to vary smoothly from point to point, having its maximum at the centre of the pattern. Frequently, however, a network of light and dark lines of the form shown in figure 5.3 is observed, crossing the diffuse background and forming what is called (after its discoverer) a Kikuchi pattern. The lines are usually arranged in pairs with one light line and one dark line parallel to each other. If the specimen is tilted slightly the diffraction pattern changes little but the array of lines moves as though rigidly attached to the specimen. They arise because of the Bragg reflection of those electrons that have previously been diffusely scattered, and their origin can be understood with the aid of figure 5.4a. Electrons will be inelastically scattered in all directions and some of them, such as those travelling in the direction A shown in the figure, will fall on a set of atomic planes at the appropriate Bragg angle θ_B. These can then be diffracted into the direction B. In addition, those electrons travelling in direction B approach the same set of reflecting planes at the Bragg angle but from the other side, and may be diffracted into direction A. The net effect of this is to tend to equalise the diffuse intensities travelling in these two directions and hence, as shown schematically in figure 5.4b, the directions A and B will appear dark and light respectively relative to their local backgrounds.

Those directions that make an angle θ_B with a plane of atoms lie on the surface of the double cone shown in figure 5.4c. Because the Bragg angle is small the intersection of such a cone with the plane of the diffraction pattern appears as two parallel straight lines. One of them is lighter than the diffuse background locally and one darker, and they are separated by a distance corresponding to $2\theta_B$. The trace on the diffraction pattern of the diffracting planes themselves lies midway between these two Kikuchi lines. Higher-order reflections from the same set of planes produce additional parallel lines outside the central pair with spacings equal to twice, three times, etc., that of the central lines. If the crystal is tilted then the traces of the different possible sets of diffracting planes move across the diffraction pattern, and therefore the Kikuchi pattern also moves. By contrast, the diffraction spots, while they may change in intensity, remain in the same positions relative to the central spot and so do not move. Kikuchi patterns are observed only when there is sufficient scattering to

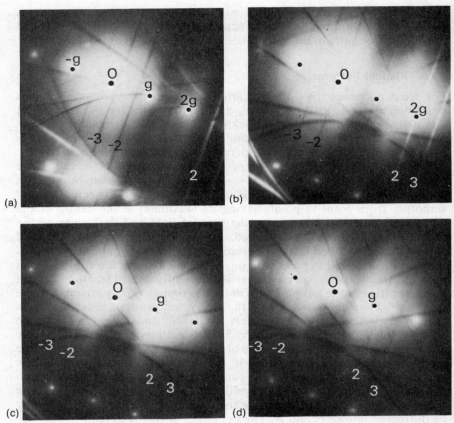

Figure 5.3. Kikuchi patterns obtained from an aluminium specimen (where spots in the diffraction pattern are obscured by the diffuse background their positions are indicated by black discs): (a) the crystal is tilted away from the reflecting position for the spot g; (b) the orientation is half-way between (a) and (c), where the crystal is almost exactly at the Bragg orientation. It can be seen that the lines labelled 3 and −3, due to the 3g and −3g reflections, almost pass through the 2g and −g spots. In (d) the specimen is on the other side of the reflecting position, half-way to the orientation where the g and −g reflections are equally strong

give an observable pattern, and when the crystallographic orientation throughout the illuminated area of the specimen is sufficiently constant. If the specimen is bent, lines produced in one area do not coincide with lines produced in another, and the pattern becomes smeared out.

Kikuchi lines, because their positions are related to the crystal planes producing them, are extremely useful for determining the exact orientation of the specimen relative to the incident beam. Thus, referring to figure 5.4a, if the specimen were tilted through the angle φ the incident

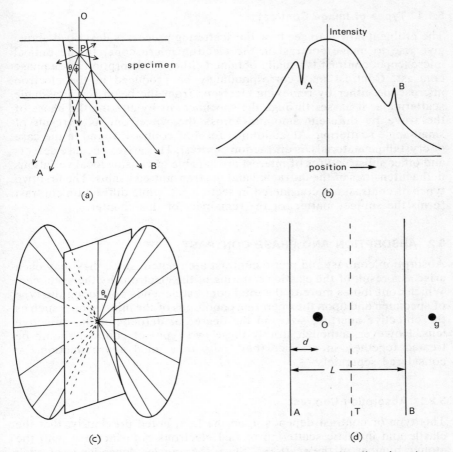

Figure 5.4. (a) Electrons scattered inelastically into the direction A are subsequently diffracted into the direction B and vice versa; (b) this tends to equalise the intensities in these two directions, making A, which is closer to the incident beam direction, dark relative to its local background and B light; (c) all directions that make an angle θ_B with the Bragg planes lie on the surface of a double cone, semi-angle $\pi/2 - \theta_B$; (d) schematic representation of the diffraction pattern corresponding to (a): L corresponds to $2\theta_B$ and d to ϕ. Thus ϕ is equal to $2\theta_B d/L$

beam would travel along the direction A and would fall on the lattice planes at the exact Bragg angle. The deviation from the exact Bragg angle is therefore φ, and this angle can be determined from the diffraction pattern, as shown in figure 5.4d. At the exact Bragg position φ is zero and one Kikuchi line passes through the diffraction spot and one through the spot at the centre of the pattern. If φ is found for two sets of reflecting planes, whose traces are not parallel to each other on the diffraction pattern, the orientation is determined.

5.1.4 Types of Image Contrast

The problem now is to see how the scattering processes described above give rise to image contrast in the electron microscope. In the optical microscope, contrast is usually obtained either as absorption or as phase contrast. Contrast may correspondingly be produced in the electron microscope either by removing electrons from the beam by large-angle scattering as it passes through the specimen, or by altering the phase of the wave by different amounts across the wavefront as a result of small-angle scattering. An additional form of contrast, found in the case of crystalline materials, is diffraction contrast. This arises because defects and other substructures of interest are capable of causing local variations in the intensities of the diffracted and the transmitted beams. The first two types of contrast are considered in section 5.2, while diffraction contrast forms the subject matter for the remainder of this chapter.

5.2 ABSORPTION AND PHASE CONTRAST

Absorption contrast and phase contrast are related, since they may each arise as a result of the elastic scattering of the electrons by the specimen. Which contributes more to the total contrast depends both upon the type of specimen and upon the operating conditions of the microscope, such as the objective aperture size and the degree of defocus of the objective lens. However, although strictly these two types of contrast should be treated together, in practice they may, to a good approximation, be considered separately.

5.2.1 Absorption Contrast

This type of contrast depends upon the fact, noted previously, that the elastic and inelastic scattering of fast electrons both increase with the atomic number of the scatterer. Since the angular dependence of such scattering is roughly the same for all atoms, heavy atoms will scatter more electrons outside the range of angles admitted by the objective lens aperture than will lighter atoms. Thus parts of a specimen that are thicker or that are composed of elements of higher atomic number will deplete the transmitted beam more and consequently appear darker than the other regions in the resultant image. The contrast observed in figure 4.12 is of this type.

The magnitude of the contrast that can be obtained in this manner depends upon the difference in 'absorbing' (that is, scattering) powers of neighbouring areas. While this can be affected slightly by changing the size of the objective aperture, it is mainly determined by the atomic numbers of the materials present and their thicknesses. A specimen in which adjacent areas are composed of materials of similar atomic numbers will thus show only low contrast. The contrast may be enhanced if a thicker specimen is used, since the difference in transmitted intensity between adjacent areas is then increased, but with a decrease in total intensity and also resolution. Alternatively, strongly scattering material can be added to some parts of the specimen but not to others. This occurs when a material is shadowed in the manner described in section 4.4, and also when biological materials are stained by being soaked in a solution of a salt of a heavy metal. Some parts of the specimen, (for example, the walls of a cell) take up the salt preferentially and consequently appear darker in the resulting micrographs than the neighbouring regions, which contain only carbon compounds and so scatter the electrons less strongly. Examples of micrographs of stained sections are shown in figure 5.5.

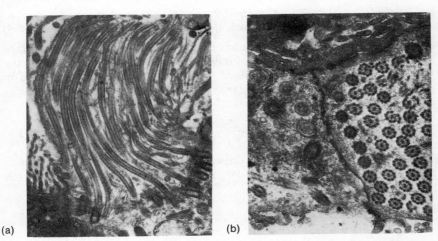

(a)   (b)

Figure 5.5. (a) Longitudinal and (b) transverse sections of respiratory columnar cells in bovine olefactory mucosa, stained by soaking in lead citrate solution with intermediate rinses in uranyl acetate solution. The preferential uptake of the heavy metal ions and the consequent increase in contrast of parts of the structure is evident in both sections. (Courtesy of B. P. Menco)

Since adequate contrast is obtained from absorption only when there is an appreciable amount of scattering and consequently many of the electrons have lost some energy, the resulting resolution tends to be low, being limited by chromatic rather than spherical aberration. The highest resolution is obtained from a conventional electron microscope only when using very thin (< 10 nm) specimens, in which case absorption is weak and an alternative contrast mechanism, namely phase contrast, has to be used.

5.2.2 Phase Contrast

Absorption contrast is due to those electrons that are scattered outside the objective aperture, but since most of the elastic scattering occurs at small angles, an even larger number of scattered electrons passes through this aperture. Since these electrons correspond to a variation in the phase of the wave leaving the lower surface of the specimen, rather than to a variation in the real amplitude, the intensity of the wave is everywhere the same, and direct imaging of the specimen produces no contrast. Thus consider a thin slice of specimen material normal to the incident beam and of thickness dz. The incident electrons are represented by a plane wave, amplitude A_0, whose wave vector in free space is χ_0. An electron in this wave at r, where the potential is $-V(r)$, has its kinetic energy increased by $eV(r)$, and there is a corresponding local change in the associated wave vector to $\chi(r)$. Since $eV(r)$ is small compared with the initial electron energy, to a good approximation $\chi(r)$ is given by

$$\chi(r) = [\chi_0^2 + 2meV(r)/h^2]^{\frac{1}{2}}$$
$$\approx \chi_0 + meV(r)/h^2\chi_0 \qquad (5.4)$$

Because the wave vector depends upon $V(r)$, which varies from place to place, the phase of the wave emerging from the lower surface of the thin slice will also vary from place to place in the manner indicated in figure 5.6. The phase variation $\delta\varphi(r)$ is given by the product of the change in wave vector and the thickness of the slice, multiplied by 2π

$$\delta\varphi(r) = \frac{2\pi me}{h^2\chi_0} V(r) \, \mathrm{d}z \qquad (5.5)$$

For an object in which the potential varies with depth the total phase shift at the lower surface can be found by integrating this expression along the

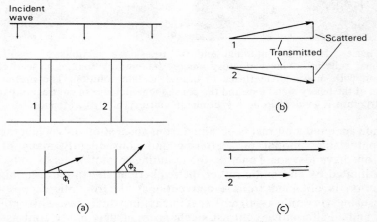

Figure 5.6. (a) A plane wave (constant amplitude and phase) is incident upon a thin layer. The average potential in column 1 is different from that in column 2 and hence the wave emerging from the lower surface, while having the same amplitude everywhere, has a different phase φ_1 at 1 from the phase φ_2 at 2. This is indicated in the phase diagrams below the layer. (b) The total wave can be regarded as a transmitted wave combined with a relatively weak set of scattered waves whose phases differ from that of the transmitted wave by $\pi/2$. As the amplitude of the scattered component varies, so does the phase of the total wave. (c) Changing the phase of the scattered component by $(n + \frac{1}{2})\pi$ converts the phase variation to an amplitude, and hence intensity difference

path travelled by any particular electron, or, equivalently, replacing $V(r)$ by its average value along that path. This average value will depend upon where the path passes through the specimen, and will be a function of x and y.

The wave at the lower surface can thus be written

$$\psi(r) = A_0 \exp(2\pi i \chi_0 \cdot r) \exp(2\pi i m e V(r) \, dz / h^2 \chi_0) \tag{5.6}$$

This wave is one that, as already described, has a constant amplitude and a variation in phase across its wavefront, and will not directly give contrast. That this phase variation corresponds to scattered electrons can be seen by assuming that it is small and expanding the second exponential to give

$$\psi(r) = A_0 \exp(2\pi i \chi_0 \cdot r)(1 + 2\pi i m e V(r) \, dz / h^2 \chi_0) \tag{5.7}$$

$V(r)$ varies in a manner determined by the structure of the specimen and in the case of a crystalline material, where $V(r)$ is periodic with the crystal periodicity, it can be written as a Fourier series. In the more general case of an amorphous material, a wide range of periodicities is present and $V(r)$ can be represented by a Fourier integral

$$V(r) = V_0 + \int_q V(q) \exp(2\pi i q \cdot r) \, dq \tag{5.8}$$

so that conversely

$$V(q) = \int_r V(r) \exp(-2\pi i q \cdot r) \, d^3 r \tag{5.9}$$

In this approximation q is effectively confined to a plane normal to the electron beam. Combining equation 5.8 with 5.7 enables the emerging wave to be written

$$\psi(r) = \psi_t + \psi_d$$

where

$$\psi_t = A_0 \exp(2\pi i \chi_0 \cdot r)(1 + 2\pi i m e V_0 \, dz / h^2 \chi_0) \tag{5.10}$$

This is a plane wave travelling in the incident direction with the incident amplitude A_0 but with a phase which differs from that of the incident wave due to the average potential V_0 of the specimen. It is effectively the incident wave with its phase retarded due to the mean refractive index of the sample.

The remainder, ψ_d, is given by

$$\psi_d = A_0(2\pi i m e / h^2 \chi_0) \int_q V(q) \exp(2\pi i q \cdot r) \, dq \, \exp(2\pi i \chi_0 \cdot r) \, dz$$

$$= A_0(2\pi i m e / h^2 \chi_0) \int_q V(q) \exp(2\pi i (\chi_0 + q) \cdot r) \, dq \, dz \tag{5.11}$$

ψ_d is seen to consist of a spectrum of waves combined together, a typical component being a plane wave with wave vector $\chi_0 + q$ and having an amplitude of $2\pi i A_0 m e V(q) / h^2 \chi_0$ per unit range of wave vector. Thus the Fourier component of the specimen potential which has a wavelength of $1/|q|$ gives rise to a scattered wave which has a wave vector $\chi_0 + q$, an amplitude proportional to A_0, to $V(q)$ and to dz, and which has a phase differing from that of the transmitted wave by $\pi/2$.

In the equivalent situation in the optical microscope, contrast is obtained by changing the phases of these scattered waves, usually by inserting a phase plate into the system as indicated in figure 5.7. The unscattered light is brought to a focus in the centre of the back focal plane

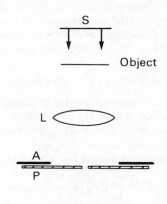

Figure 5.7. The production of optical phase contrast. The object is illuminated with plane light waves S: the phases of the scattered rays are changed by a phase plate P, positioned near the aperture A in the back focal plane of the lens L, to give contrast in the image

of the objective lens, where it passes through the hole in the phase plate and so does not suffer any change in phase. The scattered waves are focused at other points in this plane, pass through the plate and have their phases changed, normally by $\pi/2$. If the relative contributions to adjacent points in the image from the scattered waves are $\pm i\delta$ in the absence of a phase plate (where δ is small) the corresponding intensities are $|1 \pm i\delta|^2 = 1 + \delta^2$, and there is no contrast: this is shown in figure 5.6b. If the phase of the scattered contribution is changed by $\pi/2$ as indicated in figure 5.6c this becomes $|1 \pm \delta|^2 = 1 \pm 2\delta + \delta^2$, which when δ is small is approximately $1 \pm 2\delta$. The intensities at the two image points now differ by 4δ, so that changing the phases of the scattered waves produces contrast.

The prospects for using a phase plate in an electron microscope have been reviewed by Thon (1971): at the present it would not be feasible to employ such a device on a routine basis. In practice, it is possible to obtain some phase contrast using an unmodified instrument, since the phases of the scattered waves are also affected by aberrations and the degree of defocus of the objective lens. A wave scattered at an angle θ by the specimen has its phase at the image plane altered by an amount γ due to the effects of defocus and spherical aberration, where γ is given by

$$\gamma = \frac{\pi c_s \theta^4}{2\lambda} - \frac{\pi \theta^2 \Delta L}{\lambda} \qquad (5.12)$$

ΔL is the distance of the specimen below the plane on which the objective lens is focused and c_s is the spherical aberration coefficient of the lens. Since this shift in phase is a function of θ it cannot act like a simple phase plate. Its rapid variation with θ is evident from figure 5.8, where γ is plotted for a typical c_s of 2 mm and a range of values of ΔL. It can be seen that a defocus of about 90 nm would cause the shift in phase to be close to the optimum value of $-\pi/2$ for those electrons scattered through angles in the range 4×10^{-3} to 9×10^{-3} radians. These angles correspond to periodicities with wavelengths between about 0.9 nm and

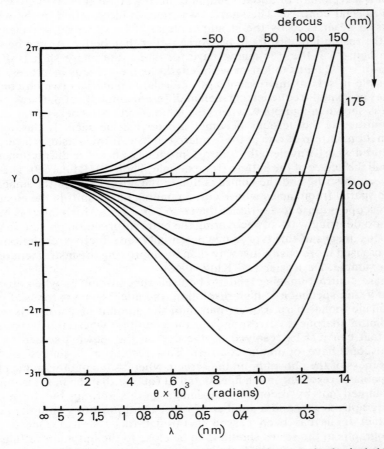

Figure 5.8. Values of γ, the phase shift due to defocus and spherical abberration, as a function of scattering angle θ for different amounts of defocus, $c_s = 2$ mm

0.4 nm, and since the corresponding electron waves are phase-shifted by about the optimum amount, detail in the object of this periodicity will be visible with near maximum contrast. Other periodicities will be visible with lower or no contrast, and inclusion of electrons scattered through angles larger than those with about the optimum phase shift will merely lower the overall contrast without increasing the resolution. If the defocus is increased still further, the magnitude of the phase shift increases until it reaches a value of $-3\pi/2$, at a defocus of about 150 nm. At this point, frequencies in the range 0.55 to 0.35 nm have maximum contrast, but contrast which is reversed compared with that given by a phase shift of $-\pi/2$. These spatial frequencies are a little shorter than those in contrast at 90 nm, and in principle the resolution should be higher. However, at this defocus there is also a range of spatial frequencies in the specimen for which the phase shift is close to $-\pi/2$, centred upon a wavelength of about 1 nm, and these will appear relatively strong with normal contrast. The shorter wavelength information in the image tends to be masked by the strong contrast of these longer wavelengths and the resolution appears to be worse rather than better. As can be seen from figure 5.7, the frequency range for which the phase shift is of the order of $-\pi/2$ tends to longer wavelengths as the defocus increases. This is visible in sets of micrographs of amorphous materials (which contain ranges of spatial frequencies) taken at different amounts of defocus. Such a set is shown in figure 5.9, where the increase in coarseness of the detail in contrast as the defocus becomes greater can be seen. If the optical diffraction pattern of one of these micrographs is formed using a laser, the obscured short-wavelength contrast produces an extra diffraction ring around the centre of the pattern (for example, Thon, 1971, figure 9). The gap between this ring and the centre is due to the absence from the image of those spatial frequencies whose corresponding phase shift in the electron microscope is close to $-\pi$ rather than $-\pi/2$ or $-3\pi/2$. Much effort is being put into devising ways of recovering the extra information present in out-of-focus images of this type, since potentially the effective resolution can be increased more easily this way than by improving the instrument itself (for example, De Rosier and Klug, 1968).

Phase contrast imaging tends to be used subconsciously when observing any thin specimen at high resolution: the microscope is focused until the image looks sharp, rather than until the amount of defocus is zero. Maximum sharpness corresponds to the condition where those frequencies that can just be resolved by the eye on the viewing screen of the microscope have optimum contrast. This resolution is limited by the coarseness of the phosphor on the screen, whereas the photographic plate is capable of resolving much finer detail. It follows that the best resolution is obtained not by recording the image at this setting, but by taking photographs at this and a number of other focal positions until the point of exact focus has been passed (a through-focus series). One of the micrographs in the series should then be close to the optimum setting for the best resolution of which the microscope is capable. This is a much more satisfactory procedure than taking micrographs of different areas at random and hoping one or more will be 'sharp'.

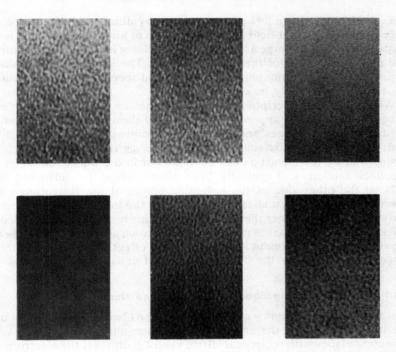

Figure 5.9. A through-focus series of micrographs of a carbon film, the change in focus between one micrograph and the next being approximately 200 nm. The increasing coarseness of the detail having strong contrast with increase in defocus is apparent; note also that the contrast of some features reverses on going through focus

5.3 DIFFRACTION CONTRAST

The two mechanisms of contrast production considered in the previous section make use of electrons scattered at relatively large angles in order to form the image, and so are capable of giving high resolution. However, many of the structures that are of interest in crystalline solids do not show up with high contrast using these methods. Dislocations, for example, do not give absorption contrast, and while they may be seen if the planes of the crystal lattice are resolved, this requires a specimen so thin that most of the dislocations originally present slip out of the sample through the surfaces. However, such defects can be made visible in quite thick specimens by using the fact that locally they distort the planes of the lattice, and thereby change the strength with which those planes diffract an incident electron wave. Contrast is produced by allowing only either the transmitted beam or one diffracted beam to pass through the objective aperture, and so form the image. Thus suppose there is a region of the specimen in which the lattice planes are bent so as to increase the diffracted intensity locally. In an image formed using the diffracted beam the distorted region will appear light compared with the surrounding areas. Conversely, since the transmitted beam will locally have lost more electrons by diffraction that region will appear dark in a bright-field image. The width of an image produced in this way depends on how large a region of the specimen is appreciably distorted, and may typically be of the order of 10 nm. The instrumental resolution required is therefore not particularly high and specimens several hundred nm thick may usefully be examined.

While a simple description of this type enables the basic principles of image formation to be appreciated, it does not allow such contrast features as stacking-fault fringes and black–white contrast at dislocations to be understood. The elastic scattering near a Bragg reflection is sufficiently strong that all the electrons can be scattered into a diffracted beam in a specimen thickness of typically 20 or 30 nm. Since the diffracted beam falls on the other side of the reflecting planes at the Bragg angle (see figure 5.10a) it is also scattered strongly into the incident beam direction. Only for crystals thinner than a few nm can this rescattering be neglected, and so a theory which takes it into account (known as the dynamical theory) is essential. This treatment also includes the effect of inelastic scattering, since this can modify the observed contrast considerably.

5.3.1 Derivation of the Equations of the Dynamical Theory

It is assumed that a plane wave of the form $\exp(2\pi i \chi_0 \cdot r)$ is incident upon the upper surface of the crystalline specimen and that only one set of reflecting planes with reciprocal lattice vector g diffracts the electrons (the two-beam approximation).

The wave vector of the diffracted wave χ is found using the Ewald sphere

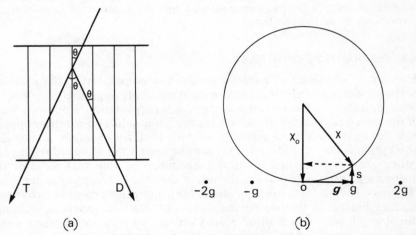

(a)   (b)

Figure 5.10. (a) When the transmitted wave T meets the reflecting planes at the Bragg angle the resultant diffracted wave D also strikes these planes at θ_B; (b) the Ewald sphere construction. The deviation parameter s for scattering from the incident to the diffracted wave has the opposite sign from that for scattering in the reverse direction

construction shown in figure 5.10b. The difference between χ and $\chi_0 + g$, the length s, is called the deviation parameter, since it describes how far the crystal is tilted away from the exact Bragg condition. Conventionally s is taken to be positive when the reciprocal lattice point lies within the sphere.

At any depth z in the crystal the electrons are represented by a wave, amplitude $\varphi_0(z)$, travelling in the incident direction and an associated diffracted wave, amplitude $\varphi_g(z)$. Since the Bragg angle is very small these two waves can be treated as travelling together down a column in the crystal, normal to g. This column is similar to the ones shown in figure 5.6a. In most cases of interest the transfer of electrons between one column and the next can be neglected (for example, Howie and Sworn 1970). Initially, at the upper surface of the crystal all the intensity will be in the incident direction and $\varphi_g(0) = 0$. As the incident wave penetrates the crystal $\varphi_0(z)$ starts to decrease and $\varphi_g(z)$ increases as it gains electrons from the incident beam. The effect of the scattering from the diffracted wave back into the incident wave then becomes important, and in general in any thin section, thickness dz, of the column, the diffracted wave will both lose electrons to the transmitted wave and gain electrons from it. The equations expressing this situation mathematically can be written down using the results of the treatment given in section 5.2, where the scattering of electrons by a non-crystalline material was considered. In the present case, since the potential within the crystal is periodic, $V(r)$ can be written as a Fourier series

$$V(r) = \sum_h V_h \exp(2\pi i h \cdot r) \tag{5.13}$$

Since it is assumed that there is only one strong diffracted beam only those terms with $h = 0, g$ are considered, in which case equation 5.11 gives the wave scattered into the diffracted wave direction by the thin slice dz as

$$\varphi_0(z)(2\pi i m e \, dz/h^2\chi_0)V_g \exp[2\pi i(\chi_0 + g) \cdot r]$$

This is a contribution to the diffracted wave, wave vector χ, already travelling in this direction. Calling this contribution $d\varphi_g \exp(2\pi i \chi \cdot r)$ we have

$$d\varphi_g = \varphi_0(z) \, dz(2\pi i m e/h^2\chi_0)V_g \exp(2\pi i(\chi_0 - \chi + g) \cdot r)$$
$$= \frac{i\pi}{\xi_g} \varphi_0 \, dz \exp(-2\pi i s \cdot r) \tag{5.14}$$

where $\xi_g = h^2\chi_0/2meV_g$. ξ_g is a length, typically of the order of 10–50 nm, called the extinction distance. It describes the strength of the Bragg scattering in the crystal, ξ_g decreasing as the strength of the scattering and hence V_g increases. Since V_g and the atomic scattering factor are both related to the atomic potential (section 5.1.1) it can be shown that

$$V_g = \frac{h^2}{2\pi me\Omega} \sum_i f_i(g) \exp(2\pi i g \cdot r_i)$$

Ω is the volume of the unit cell, and $f_i(g)$ is the scattering factor for the ith atom in the cell, located at r_i.

If we generalise equation 5.14 to include not only the scattering from the transmitted wave but also, in the manner of equation 5.10, the scattering from the diffracted wave itself, it becomes

$$\frac{d\varphi_g}{dz} = \frac{i\pi}{\xi_g} \varphi_0 \exp(-2\pi isz) + \frac{i\pi}{\xi_0} \varphi_g \qquad (5.15a)$$

ξ_0 is given by the expression for ξ_g but with V_0 in place of V_g. Similarly, for the scattering into the transmitted wave from itself and from the diffracted wave we get

$$\frac{d\varphi_0}{dz} = \frac{i\pi}{\xi_g} \varphi_g \exp(2\pi isz) + \frac{i\pi}{\xi_0} \varphi_0 \qquad (5.15b)$$

In these equations $s \cdot r$ has been replaced by sz, where s is the z component of s, since only its z component (the component normal to the plane of the specimen) is important. The sign of s is different in the two equations since, as can be seen in figure 5.10b, if s has one sign for scattering from the direct to the diffracted wave, it must have the opposite sign for scattering in the reverse direction.

These equations apply simultaneously and describe the transfer of electrons back and forth between the incident and diffracted beams for the case of a perfect crystal. In order to obtain equations that may be used to calculate the contrast from an imperfect crystal, a form of $V(r)$ appropriate to a lattice containing a defect is required in place of that given by equation 5.13. If the displacement at depth z in a given column is $\mathbf{R}(z)$ relative to the perfect crystal then the potential at r in the actual crystal is the potential which was previously, in the perfect crystal, found at $r' = r - \mathbf{R}(z)$. Hence

$$V(r) = \sum_g V_g \exp(2\pi ig \cdot r') = \sum_g V_g \exp\{2\pi ig \cdot [r - \mathbf{R}(z)]\}$$

Substituting this into equation 5.11 and proceeding as before, after changing the mean wave vector from χ_0 to $\mathbf{K}_0$ to remove the terms corresponding to the mean refractive index of the material we get

$$\frac{d\varphi_0}{dz} = \frac{i\pi}{\xi_g} \varphi_g \exp[2\pi isz + 2\pi ig \cdot \mathbf{R}(z)] \qquad (5.16a)$$

$$\frac{d\varphi_g}{dz} = \frac{i\pi}{\xi_g} \varphi_0 \exp[-2\pi isz - 2\pi ig \cdot \mathbf{R}(z)] \qquad (5.16b)$$

These equations now describe the scattering from the direct to the diffracted beam in an imperfect crystal. They are not unique, and other pairs of coupled equations suitable for particular applications may be obtained by substituting appropriately for φ_0 and φ_g. For example, the substitutions $\varphi_0' = \varphi_0$, $\varphi_g' = \varphi_g \exp[2\pi isz + 2\pi ig \cdot \mathbf{R}(z)]$, have the property that $|\varphi_0|^2 = |\varphi_0'|^2$, $|\varphi_g|^2 = |\varphi_g'|^2$, and give

$$\frac{d\varphi_0'}{dz} = \frac{i\pi}{\xi_g} \varphi_g' \qquad (5.17a)$$

$$\frac{d\varphi'_g}{dz} = \frac{i\pi}{\xi_g}\varphi'_0 + 2\pi i\left(s + g\cdot\frac{dR}{dz}\right)\varphi'_g \qquad (5.17b)$$

The validity of the simple explanation of diffraction contrast given earlier is demonstrated by these equations. s is equal to g multiplied by the appropriate component of the angular misorientation of the whole crystal, while $g\cdot(dR(z)/dz)$ is the corresponding local tilt of the reflecting planes due to the displacement field $R(z)$. The defect produces contrast by locally altering the tilt of the reflecting planes and hence changing the strength of the scattering between the transmitted and the diffracted waves.

While this simple picture may help in deciding whether contrast is likely to be strong, weak or zero under given conditions, solutions to the coupled equations are required for a complete interpretation. In almost all cases of interest, because of the form of $R(z)$, these solutions have to be obtained numerically. A few simple cases can be solved analytically and these are treated first, since an understanding of how contrast arises in these instances is of help when considering the results of numerical computations for more complicated strain fields.

5.3.2 Solutions for a Perfect Crystal

For a perfect crystal $R(z)$ is zero and hence dR/dz is also zero everywhere. Dropping the primes, equations 5.17a, b become

$$\frac{d\varphi_0}{dz} = \frac{i\pi}{\xi_g}\varphi_g \qquad (5.18a)$$

$$\frac{d\varphi_g}{dz} = \frac{i\pi}{\xi_g}\varphi_0 + 2\pi is\varphi_g \qquad (5.18b)$$

Differentiating the first of these and substituting in the second leads to

$$\frac{d^2\varphi_0}{dz^2} - 2\pi is\frac{d\varphi_0}{dz} + \frac{\pi^2}{\xi_g^2}\varphi_0 = 0 \qquad (5.19)$$

This equation has a solution of the form $\exp(2\pi i\gamma z)$ where

$$\gamma = \tfrac{1}{2}[s \pm (s^2 + 1/\xi_g^2)^{\frac{1}{2}}] \qquad (5.20)$$

Thus there are two possible values for γ, and hence two solutions for φ_0. Making the substitution

$$s\xi_g = \cot\beta \qquad (5.21)$$

we have

$$\gamma^{(1)} = -(\tan\beta/2)/2\xi_g \qquad \gamma^{(2)} = (\cot\beta/2)/2\xi_g \qquad (5.22)$$

Thus the solutions are

$$\varphi_0^{(1)} = C_0^{(1)}\exp[-i\pi z(\tan\beta/2)/\xi_g] \qquad (5.23a)$$

$$\varphi_0^{(2)} = C_0^{(2)}\exp[i\pi z(\cot\beta/2)/\xi_g] \qquad (5.23b)$$

These two independent solutions have been labelled by superscripts 1, 2 and the complete solution for the transmitted amplitude is given by their sum. The relative amplitudes $C_0^{(1)}$ and $C_0^{(2)}$ have yet to be determined. By substituting these solutions separately into equation 5.18a corresponding solutions are obtained for φ_g

$$\varphi_g^{(1)} = C_g^{(1)} \exp(2\pi i \gamma^{(1)} z) \tag{5.24a}$$

$$\varphi_g^{(2)} = C_g^{(2)} \exp(2\pi i \gamma^{(2)} z) \tag{5.24b}$$

where

$$C_g^{(1)}/C_0^{(1)} = -\tan \beta/2 \qquad C_g^{(2)}/C_0^{(2)} = \cot \beta/2 \tag{5.25}$$

The total diffracted wave amplitude is similarly the sum of these two independent diffracted wave amplitudes.

We now have to find $C_0^{(1)}$ and $C_0^{(2)}$, which are determined by the boundary conditions at the entrance surface of the crystal. Since the refractive index of the specimen is almost unity the probability of reflection at the upper surface of the specimen at or near normal incidence is extremely small. Hence the incident wave may be matched directly to a combination of the solutions found above. Taking the amplitude in the incident direction to be one at the crystal surface, at which point the diffracted-wave amplitude is zero, we have

$$C_0^{(1)} + C_0^{(2)} = 1 \qquad C_g^{(1)} + C_g^{(2)} = 0 \tag{5.26}$$

Using the relations of equation 5.25 we find

$$C_0^{(1)} = \cos^2 \beta/2 \qquad C_0^{(2)} = \sin^2 \beta/2 \qquad C_g^{(1)} = -C_g^{(2)} = -\cos \beta/2 \sin \beta/2 \tag{5.27}$$

At the exit surface of the crystal reflection is again neglected, and the total diffracted wave inside the crystal is matched to a plane wave outside the crystal travelling in the diffracted wave direction. A similar matching is carried out for the transmitted wave, so that the transmitted and diffracted intensities are equal to the intensities of the transmitted and diffracted waves just inside the crystal at the exit surface.

All the unknowns have now been found, and the diffracted intensity I_g from a perfect crystal is therefore

$$\begin{aligned} I_g &= |\varphi_g|^2 = |\cos \beta/2 \sin \beta/2 [\exp(2\pi i \gamma^{(1)} z) - \exp(2\pi i \gamma^{(2)} z)]|^2 \\ &= 4 \cos^2 \beta/2 \sin^2 \beta/2 \sin^2 [\pi(\gamma^{(1)} - \gamma^{(2)})z] \\ &= \sin^2 \beta \sin^2 (\pi t/\xi_g \sin \beta) \end{aligned} \tag{5.28}$$

Since inelastic scattering, which removes electrons from the diffracted and transmitted waves, has not yet been included in the theory the total number of electrons is constant and the transmitted intensity I_0 is given by

$$I_0 = |\varphi_0|^2 = 1 - I_g$$

These intensities depend both upon the thickness t of the specimen and (through the parameter β) upon its orientation. At the exact Bragg orientation $\beta = \pi/2$, $\sin \beta = 1$ and the diffracted intensity is simply

$$I_g = \sin^2 (\pi t/\xi_g) \tag{5.29}$$

For crystals for which $t \ll \xi_g$ this is approximately $(\pi t / \xi_g)^2$, so that the diffracted intensity is proportional to t^2, rather than to t as might be expected. However (as in the case of X-rays), the angular range over which the reflected intensity is appreciable is proportional to $1/t$. Consequently the integrated intensity, which is the quantity measured in conventional diffraction experiments, and which depends both on the peak intensity and on the range of reflection, is proportional to t.

As t increases the diffracted intensity does not increase indefinitely, but reaches a maximum value of unity at a crystal thickness of $\xi_g/2$, at which point the transmitted beam intensity is zero. The situation is now identical to that just inside the entrance surface of the specimen, except that the diffracted and transmitted wave directions are interchanged. Electrons are now scattered back into the transmitted beam direction so that after the waves have travelled a further distance of $\xi_g/2$ through the crystal the diffracted intensity has fallen to zero and the electrons are once again all in the transmitted beam. This cycle repeats itself in each successive layer of thickness ξ_g (see figure 5.11) and may be observed experimentally using a

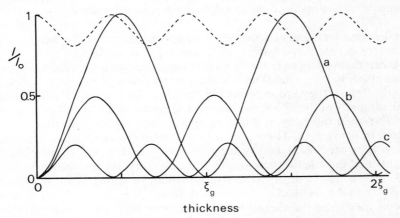

Figure 5.11. The intensity I (in terms of the incident intensity I_0) in the diffracted wave from a perfect crystal, neglecting absorption, as a function of thickness at (a) $s = 0$, (b) $s = 1/\xi_g$, and (c) $s = 2/\xi_g$. The dashed curve is the transmitted intensity for $s = 2/\xi_g$, and is complementary to curve c

specimen consisting of a wedge-shaped crystal. When the crystal is set at the Bragg angle for a low-order reflection, fringes parallel to the edge of the specimen are visible in both the bright- and dark-field images. In figure 5.12a, which is a bright-field micrograph of such a specimen, these fringes can be seen, each light fringe corresponding to a specimen thickness at which the transmitted wave intensity is high, and each dark fringe to one at which it is low. Successive fringes of the same type correspond to thickness increments of ξ_g, and the peaks of the light fringes appear at thicknesses of ξ_g, $2\xi_g$, $3\xi_g$, etc., so that these fringes provide a means of determining the crystal thickness. A dark-field micrograph of the same area is shown in figure 5.12b,

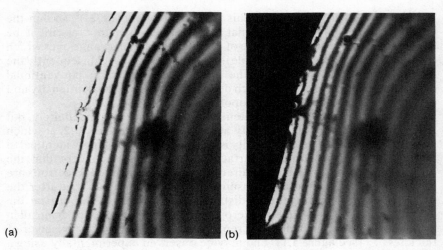

(a)   (b)

Figure 5.12. (a) Bright- and (b) dark-field micrographs of a wedge-shaped aluminium crystal, taken close to a Bragg reflecting orientation

and the complementary nature of the fringes predicted by the theory can be seen. Some calculated values of ξ_g are listed in table 5.1.

At orientations away from the exact Bragg position β is no longer $\pi/2$, but tends either to 0 or π, and the thickness fringes described above are closer together, since they are separated in thickness by $\xi_g \sin \beta$. In addition, the peak intensity of the diffracted beam is reduced by a factor of $\sin^2 \beta$, so that the diffracted intensity at different orientations varies with thickness as shown in figure 5.11. The contrast from a bent, wedge-shaped crystal, in which both thickness and orientation vary, is thus quite complicated. Bright- and dark-field micrographs of such a crystal are shown in figure 5.13, where the form of these fringes which lie along lines of constant $\sin \beta / t$ can be seen. The fringe pattern at A in figure 5.13a is more complicated than the theory predicts since in this region the reflections corresponding to the reciprocal lattice vectors g and $-g$ are both operating. Even more complex patterns occur at orientations of high

Table 5.1. Extinction distances in nm

Reflection	Al	Cu	Nb	Ge	Diamond
110	—	—	26	—	—
111	56	24	—	43	48
200	67	28	37	—	—
220	106	42	54	46	66
310	—	—	62	—	—
311	130	50	—	77	124
222	139	53	70	—	—

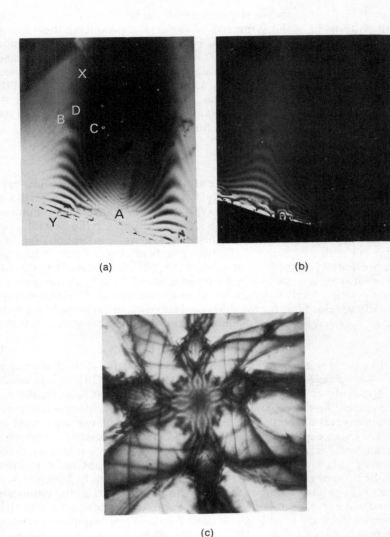

(a)   (b)

(c)

Figure 5.13. (a) Bright-field micrograph of a bent wedge-shaped crystal. The planes along the line X–Y are at an exact Bragg orientation, and thickness fringes with the maximum spacing of ξ_g are visible along this line; (b) is the corresponding dark-field micrograph. The decreasing separation of the fringes with deviation from the reflecting position can be seen. Note the symmetry of the contour in the dark field compared with the asymmetric bright-field image. (c) Complex pattern of fringes and bend contours observed in a nickel foil bent to a saucer-shape. The orientation at the centre of the pattern is [100]. The symmetry in these patterns forms the basis of the technique known as real-space crystallography (Steeds, 1973)

symmetry, an example being shown in figure 5.13c. These fringe and bend contour patterns have the symmetry of the crystal producing them and so can be used to obtain crystallographic information about the material under observation (for example, Steeds, 1973). This is particularly useful in cases where large crystals suitable for X-ray crystallography cannot be obtained.

While this treatment is reasonably accurate in giving the positions of the fringes observed with perfect crystals, and forms a starting point for the calculation of image contrast from imperfect crystals, it suffers from two deficiencies which are now briefly discussed.

5.3.3 Many-beam Theory

In the treatment outlined above only one diffracted beam has been considered, whereas experimentally it is often observed, especially at orientations of high symmetry, that several reflections are strong simultaneously. While such orientations can be avoided by tilting the crystal, even under the most favourable conditions the systematic reflections (corresponding to the line of reciprocal lattice points shown in figure 5.10b) are always present. When the point at g lies on the reflecting sphere the value of s for the point at $-g$ is equal to g^2/χ. From equations 5.21 and 5.28 it is apparent that a reflection is likely to be strong when its value of $s\xi_g$ is of the order of about unity or less. Thus for the diffracted wave corresponding to $-g$ to be neglected we must have $s\xi_g$ for that reflection much greater than 1, that is

$$g^2\xi_g/\chi \gg 1 \tag{5.30}$$

Typical values for a low-order reflection in a material such as copper at 100 keV would be $\xi_g \sim 30$ nm, $g \sim 5 \times 10^9$ m^{-1}, $\chi = 27 \times 10^{10}$ m^{-1}, in which case $g^2\xi_g/\chi$ is about 3. This indicates that neglecting this reflection is likely to be a reasonable first approximation, but that for high accuracy a treatment that takes several diffracted waves into account simultaneously must be employed. This is especially true for crystals of heavy elements, for which ξ_g is generally smaller, and at higher electron energies, when χ is greater. In the X-ray case $g^2\xi_g/\chi$ is of the order of 10^4 under these conditions, indicating that the two-beam approximation is extremely good for X-rays. In principle, sets of equations coupling together the transmitted wave and all the important diffracted waves can be obtained using the approach of section 5.3.1, but it is more straightforward to obtain solutions for the wave amplitudes starting with Schrödinger's equation (Hirsch *et al.*, 1965). A many-beam theory is necessary to explain many interesting effects, but is above the level of the present treatment.

5.3.4 Inelastic Scattering

The treatment given above is also inadequate in that it does not allow for the diffuse scattering of the electrons. This not only lowers the resolution by increasing the spread in energy of the electrons but also causes thicker regions of the specimen to appear darker in the final image. Simple attenuation resulting from randomly distributed scattering events would give rise to an exponential decay of the wave amplitudes with thickness. However, the situation is more complex than this, as can be seen in the micrograph in figure 5.13a. Not only does the overall intensity decrease with increasing thickness, as expected, but the fringe contrast decreases also, so that although the fringes have more or less disappeared at D, there is still sufficient transmitted intensity to enable detail in the specimen to be seen. In addition at B and C (where the thicknesses are similar but the signs of the deviation parameter s are opposite) the intensities are very different, whereas simple absorption would result in their being the same.

Contrast effects such as these can be explained if it is assumed that one of the two waves contributing to the direct or to the diffracted wave is scattered more strongly than the other. Physically why this should happen is discussed later, but to fit the observations the solution numbered 2 in section 5.3.2 has to have a larger absorption coefficient than the other, type 1 solution. This can be achieved if the inelastic scattering is taken into account by adding an imaginary part $iV'(r)$ to the crystal potential (this has been justified by Yoshioka, 1957). Just as the normal crystal potential gives rise to a change in the real part of the electron wave vector, the imaginary potential gives an imaginary component to the wave vector which, if $V'(r)$ is of the correct sign, corresponds to an attenuation of the electron waves as they travel through the specimen. (If k becomes $k + iq$, the wave $\exp(2\pi ikz)$ becomes $\exp(2\pi ikz - 2\pi qz)$, so the attenuation factor is $\exp(-2\pi qz)$.) By analogy with equation 5.12a we write

$$\xi'_g = h^2\chi_0/V'_g 2me \tag{5.31}$$

The equations 5.15a and b now become

$$\frac{d\varphi_0}{dz} = i\pi\left(\frac{1}{\xi_0} + \frac{i}{\xi'_0}\right)\varphi_0 + i\pi\left(\frac{1}{\xi_g} + \frac{i}{\xi'_g}\right)\varphi_g\exp[2\pi isz + 2\pi ig\cdot R(z)] \tag{5.32a}$$

$$\frac{d\varphi_g}{dz} = i\pi\left(\frac{1}{\xi_0} + \frac{i}{\xi'_0}\right)\varphi_g + i\pi\left(\frac{1}{\xi_g} + \frac{i}{\xi'_g}\right)\varphi_0\exp[-2\pi isz - 2\pi ig\cdot R(z)] \tag{5.32b}$$

The other coupled equations obtained from these are similarly modified. These equations can no longer be solved exactly, even for the case of a perfect crystal. However, in cases of practical importance the inelastic scattering is relatively weak and ξ'_g is of the order of ten times larger than ξ_g. By neglecting small quantities they may be solved approximately (Hirsch *et al.*, 1965, p. 206) to give two solutions as before. These have the same real wave vectors as the waves found in section 5.3.2, so that the depth periodicity of the thickness fringes is still $\xi_g \sin \beta$. The coefficients C_0 and C_g are also unaltered, so that to this order of accuracy the only difference is that each wave is multiplied by an attenuation factor $\exp(-2\pi q^{(i)}z)$, where

$$q^{(1)} = \frac{1}{2}\left(\frac{1}{\xi'_0} - \sin\beta/\xi'_g\right) \tag{5.33a}$$

$$q^{(2)} = \frac{1}{2}\left(\frac{1}{\xi'_0} + \sin\beta/\xi'_g\right) \tag{5.33b}$$

Taking ξ'_0 and ξ'_g to be positive causes $q^{(2)}$ to be larger than $q^{(1)}$, as required. In principle ξ_0' and ξ_g' can be calculated theoretically, but in practice they are found by fitting the calculated contrast of thickness fringes and simple defects such as stacking faults to experimental observations. These values can then be used when calculating the contrast from more complicated defects. Some of the inelastically scattered electrons pass through the objective aperture and contribute to the image so that in theory changing the size of this aperture alters the effective absorption and hence ξ_0' and ξ_g'. However, since the electrons scattered inelastically through small angles give contrast similar to that due to the elastically scattered electrons, the effective absorption does not change rapidly with aperture size. Values of ξ_0' and ξ_g' found using an average-sized aperture are usually adequate for all calculations. Some examples of calculated fringe profiles including absorption are shown in figure 5.14.

5.3.5 Kinematical Approximation

In the case of X-rays the scattering is sufficiently weak, except under rather special circumstances (see chapter 6), that the diffracted intensity is always relatively small and rescattering into the incident beam direction can be neglected. Early treatments of the contrast of defects in the electron microscope were based on this approach (known as the kinematical approximation) and when rescattering into the transmitted wave is not taken into account equations 5.16a and b become

$$\frac{d\varphi_g}{dz} = \varphi_0 \frac{i\pi}{\xi_g} \exp\left[-2\pi i s z - 2\pi i g \cdot \mathbf{R}(z)\right] \tag{5.34}$$

The diffracted amplitude from a crystal of thickness t is therefore

$$\varphi_g(t) = \frac{i\pi}{\xi_g} \varphi_0 \int_0^t \exp\left[-2\pi i s z - 2\pi i g \cdot \mathbf{R}(z)\right] dz$$

For a perfect crystal ($R = 0$) this can be integrated to give

$$\varphi_g(t) = \frac{i\pi t}{\xi_g} \varphi_0 \frac{\sin \pi t s}{\pi t s} \exp\left(-i\pi t s\right) \tag{5.35}$$

Taking the incident amplitude φ_0 to be unity gives

$$I_g = \left(\frac{\pi t}{\xi_g}\right)^2 \frac{\sin^2 \pi t s}{(\pi t s)^2} \tag{5.36}$$

Expressed in this form the corresponding result given by the dynamical theory (equation 5.28) is

$$I_g = \left(\frac{\pi t}{\xi_g}\right)^2 \frac{\sin^2 \left[\pi t (s^2 + 1/\xi_g^2)^{\frac{1}{2}}\right]}{(\pi t)^2 (s^2 + 1/\xi_g^2)} \tag{5.37}$$

These expressions are similar except that $s^2 + 1/\xi_g$ becomes s^2 in the kinematical approximation. This is valid only if $s \gg 1/\xi_g$, a condition met previously when considering many-beam effects, implying that the reflection is weak. When s is smaller than this the kinematical intensity

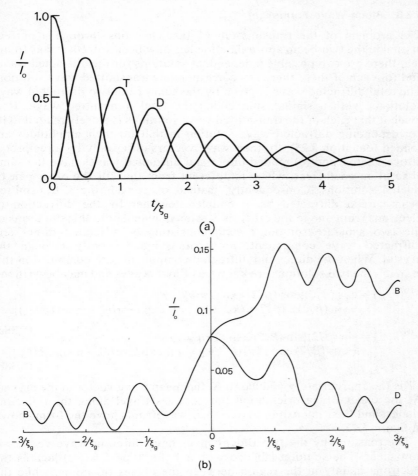

Figure 5.14. (a) Calculated thickness fringe profiles for transmitted (T) and diffracted (D) waves, including absorption: $\xi_0' = 10\xi_g$; $\xi_g' = 15\xi_g$; (b) transmitted and diffracted intensities as a function of orientation (rocking curves) for a thickness of $\frac{3}{4}\xi_g$ and the same absorption parameters. The bright-field asymmetry is evident

diverges from that given by the more exact treatment, until at the Bragg position ($s = 0$) it becomes $(\pi t/\xi_g)^2$. This does not oscillate with thickness, but is the same as the dynamical intensity provided that $t \ll \xi_g$, corresponding to the case of a crystal only a few atoms thick. The kinematical approximation has the further disadvantage that the effects of absorption cannot easily be included. Its original advantage was that calculations of defect contrast do not need as much computing time as the dynamical equations, and it is still sometimes useful in situations where the deviation from the Bragg position is large.

5.3.6 Bloch Wave Formalism

The problem of the transmission of fast electrons through a perfect crystal in the two-beam approximation has now been solved. It was found that there are two possible independent solutions for the diffracted wave, and for each of these there is a corresponding transmitted wave solution. The total diffracted wave is given by the sum of the two diffracted wave solutions, with a similar sum giving the total transmitted wave. It is evident that each of the transmitted wave solutions is closely related to its corresponding diffracted wave solution in that (a) their amplitudes are related (equation 5.25), (b) their wave vectors differ only by a reciprocal lattice vector, and (c) related direct and diffracted waves have the same absorption coefficient, which is different from that for the other pair of related solutions. Consequently, instead of considering a transmitted beam and a diffracted beam coupled together by the diffraction of electrons from one to the other, it is often convenient to think in terms of the two separate solutions, each consisting of a related direct and diffracted wave component, propagating independently through the crystal. When the direct and diffracted components are combined in this manner, the two solutions are known as Bloch waves and may be written

$$b^{(1)}(k, g) = \cos(\beta/2)\{\cos(\beta/2)\exp(2\pi i k_0^{(1)} \cdot r)$$
$$- \sin(\beta/2)\exp[2\pi i(k_0^{(1)} + g) \cdot r]\}\exp[-\pi/\xi_0' + \pi \sin(\beta/\xi_g')]$$
$$(5.38a)$$
$$b^{(2)}(k, g) = \sin(\beta/2)\{\sin(\beta/2)\exp(2\pi i k_0^{(2)} \cdot r)$$
$$+ \cos(\beta/2)\exp[2\pi i(k_0^{(2)} + g) \cdot r]\}\exp[-\pi/\xi_0' - \pi \sin(\beta/\xi_g')]$$
$$(5.38b)$$

$k_0^{(i)}$ is the appropriate $\gamma^{(i)}$ added to K, the mean wave vector in the crystal. At the exact Bragg orientation these waves travel along the reflecting planes, and their intensities across their wavefronts have the form shown in figure 5.15.

The reason why the two Bloch waves have different wave vectors is now clear. The solution labelled with a 1 (the type 1 wave) locates the electrons mainly in the region between the planes of atoms, while the other wave (type 2) concentrates them at the reflecting planes. Therefore the type 2 wave electrons will have a lower average potential energy and hence greater kinetic energy than those of type 1, and hence a larger wave vector. A similar argument explains the greater absorption of the type 2 wave as being due to the concentration of its electrons into regions near the centres of the atoms where the probability of occurrence of inelastic scattering is higher. The type 1 wave avoids these regions, so that it is less scattered and consequently has a smaller absorption coefficient. Scattering with the excitation of electrons in the inner shells of the atoms and thermal diffuse scattering are processes that are particularly well-localised and act preferentially upon the type 2 wave. Away from the exact Bragg orientation the two Bloch waves tend to look more like plane waves, the degree of localisation which they produce is consequently less, and the difference between their absorption coefficients therefore smaller, as predicted by equation 5.33.

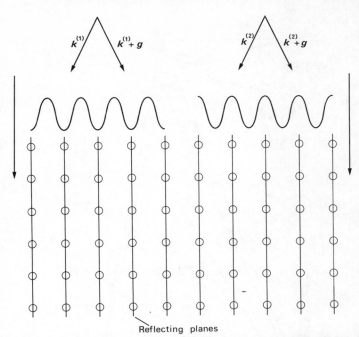

Figure 5.15. Intensities of the two Bloch waves at the reflecting position: the waves travel in a direction parallel to the reflecting planes, so that the electrons in the type 1 wave are channelled between the atoms. Those in the type 2 wave pass close to the centres of the atoms and are more likely to be scattered inelastically

It should be noted that some authors number the two waves in the opposite way so that the wave here numbered 2 becomes 1 and vice versa. This is convenient when considering many-beam situations, but the present system is the one used originally and is consistent with the numbering convention used in the case of X-rays (chapter 6).

5.3.7 Dispersion Surface Representation

A useful geometrical representation in reciprocal space of these Bloch waves and their wave vectors, known as the dispersion surface representation, is shown in figure 5.16.

The wave vector χ_0 of the wave incident upon the crystal is drawn to end at the origin of reciprocal space, O. The effect of the average crystal potential is to increase the electron wave vector by $meV_0/h^2\chi_0$ along the line normal to the crystal surface to the point P, giving a wave vector of K. Because the Bloch waves do not experience the average crystal potential, but are concentrated near to or away from the centres of the atoms, as described above, they do not have this wave vector K, but have wave vectors which differ from it by $\gamma^{(1)}$ and $\gamma^{(2)}$, and which are then represented by the points A and B. The vector AO is now the wave vector

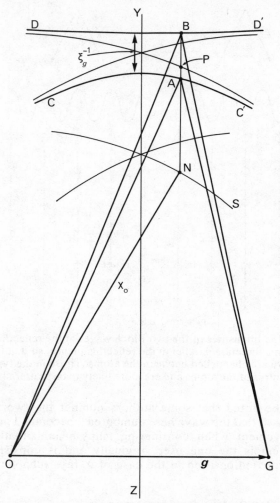

Figure 5.16. Representation of a section through the dispersion surface. If drawn to scale with g 1 cm long, then OA would be about 30 cm long and the separation of the branches of the dispersion surface, AB, would be about 30 μm

of the transmitted component of the type 1 wave, while BG is the wave vector of the diffracted component of the type 2 wave, and similarly for BO and AG.

As the direction of χ_0 and hence K, is varied in the plane of the diagram $\gamma^{(1)}$ and $\gamma^{(2)}$ change, since s changes, and the points A and B move along the lines CC' and DD', while if χ_0 is moved out of the plane of the diagram, A and B move on the surfaces generated by the rotation of CC' and DD′ about OG. The surface generated in reciprocal space in this manner is called the dispersion surface, and in the present approximation of only one diffracted beam the surface has two branches. Figure 5.16 thus shows a section through the dispersion surface: a three-dimensional sketch of the surface is shown in figure 5.17. As the crystal is tilted away from the Bragg orientation the Bloch waves tend to plane waves, and their wave vectors tend either to K, or to $K + g$. At these orientations the branches of the dispersion surface approach asymptotically spheres of radius K centred on O and G. The line OG is bisected by a plane, the Brillouin zone boundary, which cuts the dispersion surface at points which represent the two solutions at the exact Bragg orientation. The separation of these points is the reciprocal of ξ_g, the depth periodicity of the fringes at this orientation. At other orientations the separation of the two branches of the dispersion surface is $1/\xi_g \sin \beta$, corresponding to the shorter depth periodicity of the thickness fringes under these conditions. The dispersion surface representation is useful for discussing the contrast

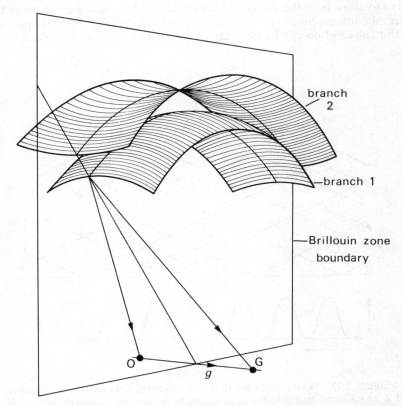

Figure 5.17. Sketch of the three-dimensional shape of the dispersion surface

from certain types of imperfections in both electron micrographs and in X-ray topographs (chapter 6).

The remainder of this chapter is devoted to the application of these ideas to various types of defect and crystal substructure. Since there is not space to consider more than a limited number of cases, a representative selection of simple examples is considered, and a brief indication given in each case of what other defects may be treated in a similar manner.

5.4 FRINGE CONTRAST FROM BOUNDARIES, FAULTS, ETC.

A number of ways in which fringes may be produced in the image is considered in this section. In all cases the fringes may be thought of as arising from the interference between two waves travelling in slightly different directions, in the manner shown in figure 5.18. The simplest fringes of this type, known as lattice fringes, are obtained from a crystal close to a reflecting orientation when the objective aperture is positioned so as to allow both the direct and the diffracted beams to pass through it to form the image. Since these fringes are only observed when the specimen is thin, absorption can be neglected, in which case the amplitude A in the

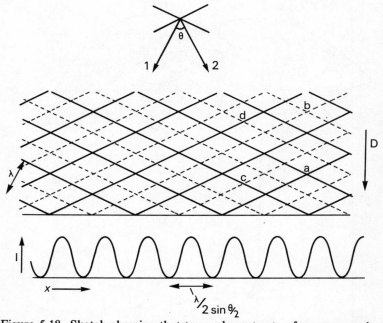

Figure 5.18. Sketch showing that two coherent sets of waves, numbered 1 and 2, crossing at an angle θ can give rise to fringes on a screen, the spacing of these fringes being $\lambda/2 \sin \theta/2$

image plane at the exact Bragg orientation is given by

$$A = [C_0^{(1)} \exp(2\pi i \gamma^{(1)} t) + C_0^{(2)} \exp(2\pi i \gamma^{(2)} t)] \exp(2\pi i K \cdot r)$$
$$+ [C_g^{(1)} \exp(2\pi i \gamma^{(1)} t) + C_g^{(2)} \exp(2\pi i \gamma^{(2)} t)] \exp(2\pi i (K+g) \cdot r]$$
$$= \cos(\pi t / \xi_g) \exp(2\pi i K \cdot r) + i \sin(\pi t / \xi_g) \exp[2\pi i (K+g) \cdot r] \quad (5.39)$$

Lattice fringes, if any, appear as the result of interference between the direct wave, represented by the first term in this expression, and the diffracted wave, represented by the second. At specimen thicknesses which are multiples of $\xi_g/2$, either the direct or the diffracted beam has zero amplitude and no interference fringes can be produced. Fringes are produced with maximum contrast when the two waves have equal amplitudes, which occurs when the thickness takes one of the values $(n + \frac{1}{2})\xi_g/2$. At a thickness of $\xi_g/4$ ($n = 0$) the amplitude is given by

$$A = (2)^{\frac{1}{2}} \cos(\pi g \cdot r + \pi/4) \exp[2\pi i (K + g/2) \cdot r] \exp(i\pi/4)$$

For convenience g may be taken to be in the x-direction, in which case the image intensity I is given by

$$I = 2 \cos^2(\pi g x + \pi/4) \quad (5.40)$$

This corresponds to a set of fringes which have the same spacing as the reflecting planes and are also aligned parallel to them. They do not, however, occur at the positions of the lattice planes, but are displaced by a quarter of an interplanar spacing in one direction. At a thickness of $3\xi_g/4$ they are displaced by an equal amount in the opposite direction. Furthermore, if the phase of the diffracted wave is altered relative to that of the transmitted wave by the effects of defocus and spherical aberration, the fringes suffer additional displacements. Therefore there is no simple relationship between the positions of the fringes and the corresponding lattice planes. However, if the crystal contains a defect such as an edge dislocation which upsets the regularity of the crystal planes, a similar dislocation appears in the fringes (figure 5.19). Consequently, obtaining these fringes is often conveniently referred to as resolving the planes of the lattice although, as pointed out above, the fringes do not in general coincide with the edges of the diffracting planes.

Fringes are produced in a similar manner by a variety of boundaries and faults in otherwise perfect crystals. Following the general approach of section 5.3 a specimen containing one of these is treated as two perfect crystals joined along an interface (which may or may not be planar, depending upon the type of boundary). The problem of the propagation of electrons in each of the crystals can be solved, and, as at the entrance surface, it is assumed that the probability of reflection at the interface can be neglected, so that the waves in the first crystal are matched at the interface to a suitable combination of waves in the second crystal. The probability of reflection at the interface is very small unless the electrons strike it at glancing incidence. The waves in the second crystal propagate

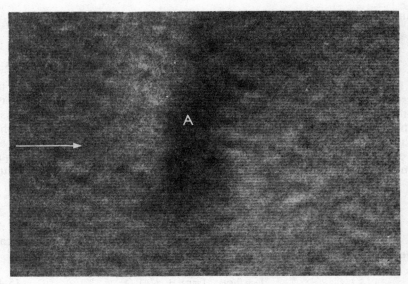

Figure 5.19. A micrograph of 0.31 nm (111) lattice plane fringes in silicon, showing a dislocation near A; the arrow indicates the extra half plane. (I. F. L. Ray, courtesy of A.E.I. Ltd)

to the exit surface where they give rise to the transmitted and diffracted waves used to form the bright- and dark-field images.

It is usually convenient to use the dispersion surface representation when treating this type of problem, and the relatively trivial case of a grain boundary is considered first as an example.

5.4.1 Grain Boundary Fringes

Figure 5.20a shows schematically a section through a specimen containing an inclined grain boundary. Grain A is close to a Bragg reflecting orientation (reciprocal lattice vector g), so that the electrons which pass through it are represented by a combination of two Bloch waves, which correspond to the points P and Q on the dispersion surface shown in figure 5.20b. Grain B is not close to any reflecting orientation and so electrons passing through it are represented by a plane wave whose wave vector has magnitude $|K|$. Apart from a slight difference in wave vector the only difference between grain B and free space is that due to inelastic scattering; grain B uniformly attenuates the waves passing through it whereas free space does not. Therefore, since a wedge-shaped single crystal gives rise to thickness fringes when close to a reflecting orientation, so also will a specimen containing an inclined grain boundary when either grain is near to a Bragg position. Because of the absorption in the second, non-diffracting grain, the intensity difference between the thick and thin areas of the reflecting grain is reduced. Figure 5.21a shows a bright-field micrograph of an inclined grain boundary which intersects the

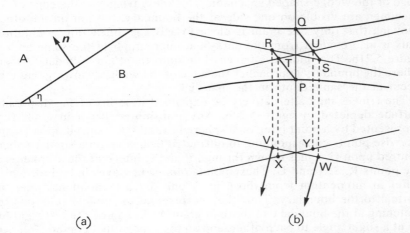

(a)   (b)

Figure 5.20. (a) Section through a specimen containing a grain boundary; (b) the dispersion surface corresponding to (a). P and Q represent the waves in the crystal above the boundary, and R, S, T and U the waves below. V, W, X and Y represent the waves outside the crystal

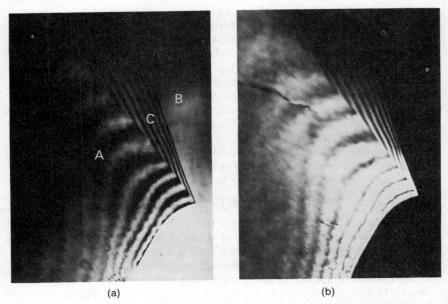

(a)   (b)

Figure 5.21. (a) Bright-field micrograph of a wedge-shaped foil containing a grain boundary. Thickness fringes occur in grain A which is close to a reflecting orientation, and these match onto the fringes in the boundary at C. (b) The dark-field micrograph corresponding to (a): the grain B is dark, showing that it is not diffracting

edge of the wedge-shaped specimen. Thickness fringes at the edge of the specimen are visible on one side of the boundary, but not on the other, showing that only one grain is close to a Bragg diffracting orientation. This is also evident from the dark-field micrograph of the same area in figure 5.21b. The thickness fringes join onto the grain boundary fringes where the boundary approaches the edge of the wedge, showing that they occur at the same depth in the material.

The fringes may alternatively be explained in terms of the dispersion surface depicted in figure 5.20b. Any transmitted beam in grain B is represented by a point lying on a sphere of radius K centred upon O, and likewise points corresponding to diffracted beams lie on a similar sphere centred upon G. Lines drawn through P and Q intersect these spheres at the points R, S, T and U. These points represent waves in grain B which differ in momentum from those in A only by a component along the normal to the boundary. Thus they correspond to suitable solutions for matching at the boundary to those in grain A. As can be seen RO and SO lie at a slight angle to each other, and so the transmitted components will interfere in the manner described previously to give fringes. In principle, these fringes exist throughout grain B and in particular at the lower surface of the specimen, where the waves in grain B match onto direct and diffracted waves outside the specimen. The waves with vectors RO and SO can be matched to a single wave outside the crystal with an intensity variation (that is, the fringes) across its wavefront. Alternatively, they may each be matched separately to plane waves represented by the points V and W on the sphere centred on O with radius χ_0 (the electron wave vector in free space). These waves interfere in the region below the specimen and so produce the fringes. These two ways of describing the fringe formation are equivalent, since two coherent plane waves travelling at a slight angle to each other are identical to a plane wave with an intensity variation across its wavefront. Because of the relatively small angle between the interfering waves, the position of the fringes is not noticeably affected by changing the microscope focus, unlike the case of lattice fringes, where this angle is large.

The angle between RO and SO is RS/K, and as RS is equal to PQ $\tan \eta = \tan \eta / \xi_g$, the fringe spacing in object coordinates is given by

$$d = \lambda / (\tan \eta / \xi_g K) = \xi_g / \tan \eta$$

This distance across the image of the boundary corresponds to an increase in depth of the boundary of $d \tan \eta$ (see figure 5.20a) which is simply ξ_g, the usual depth periodicity of thickness fringes at the Bragg position. TG and UG lie at a similar angle to each other, and fringes with the same spacing appear in the dark-field image.

While this is a convenient method of discussing the fringe spacing, since the intensities of the waves represented by R, S, T and U vary with the depth of the boundary it is less suitable as a basis for calculating the image intensity. This is more easily found by the method used in section 5.3.

Fringes of this type occur when one crystal reflects while the other does not, and the interface between them makes an angle with the plane of the specimen. Thus they may be produced at boundaries between one phase and another in a two-phase material, and at twin boundaries in a pure material, provided that one twin reflects and the other does not. When both twins reflect the fringes may be more complicated, except in the case where the reflection is from a set of planes that are continuous across the boundary which is then, of course, invisible.

5.4.2 Stacking Fault Fringes

When a crystal contains a stacking fault the lattice on one side of the fault is displaced by a constant amount relative to the lattice on the other side, so that the orientation of the crystal on either side of the fault is the same. Stacking faults lie on low-index crystallographic planes and so, unless they contain other defects such as partial dislocations, they are planar. A simple example is a stacking fault in an f.c.c. metal such as copper or silver. The structure of these crystals may be considered as being produced by stacking close-packed (111) planes in a prescribed sequence (Hull, 1965, page 16) and a fault is present if the stacking sequence is incorrect. Such an error can arise if locally a plane of atoms is removed or an extra one inserted, as occurs when vacancies or interstitials aggregate on or between the close-packed planes. In this case the crystal below the fault suffers a constant displacement R relative to the crystal above, along a direction normal to the planes, where R is $\pm a/3 \{111\}$, the sign depending upon whether the atoms are removed or added. Alternatively, one part of the crystal may be considered as having been sheared relative to the rest by a displacement vector lying in the fault plane. The possible shear vectors differ from the corresponding displacement noted above only by a vector of the perfect lattice, and so are alternative ways of describing the same faults. In general cases it may not be possible to represent R by a vector that lies in the fault plane: consequently in the real crystal, material has to be added or removed at the fault plane to ensure that the two parts join up. When a crystal containing such a fault which is inclined to the plane of the specimen is observed in either bright field or dark field using a suitable Bragg reflection, the fault appears with contrast consisting of fringes parallel to the lines of intersection of the fault with the surfaces of the specimen. Typical bright- and dark-field micrographs of a fault are shown in figure 5.22.

The intensities in the bright- and dark-field images can be calculated in a manner that is basically similar to that employed in the case of a grain boundary. The waves representing the electrons in the crystal above the fault are simply those for a perfect crystal given by equations 5.23, 5.24 and 5.27. Since the crystal orientation is the same below the fault as above it might at first be thought that the same waves would represent the electrons in this region also. Because the orientation is the same, the waves in both parts of the crystal are indeed represented by the same dispersion surface, but because the lattice is displaced the Bloch waves below the fault are also displaced and consequently are mathematically slightly different.

Since the lattice below the fault is displaced by R the potential at r is

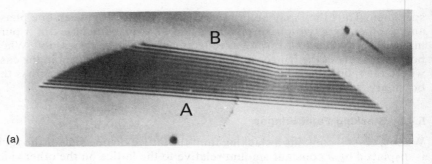

(a)

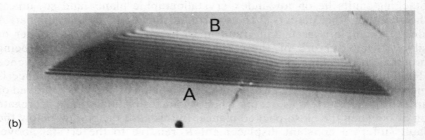

(b)

Figure 5.22. (a) Bright- and (b) dark-field micrographs of a stacking fault in VC. The fringes are similar where the fault meets the upper surface at A and complementary at B. (Billingham and Lewis; courtesy of *Phil. Mag.*)

the potential which was at $r - R$ before the displacement was introduced. Hence below the fault we have

$$V(r) = \sum_g V_g \exp\left[2\pi i g \cdot (r - R)\right] \qquad (5.41)$$

Following the method of section 5.3.1, with R a constant displacement, we find for the Bloch waves below the fault when referred to an origin in the upper part of the crystal

$$b^{(i)}(k, g) = C_0^{(i)} \exp\left(2\pi i k_0^{(i)} \cdot r\right) + C_g^{(i)} \exp\left(2\pi i (k_0^{(i)} + g) \cdot r\right) \exp\left(2\pi i g \cdot R\right) \qquad (5.42)$$

Thus the Bloch waves below the fault differ from those above in that the diffracted components are multiplied by $\exp(i\alpha)$ where α, known as the phase angle of the fault, is $2\pi g \cdot R$.

The waves that are to be used in the crystal below the fault can now be found with the aid of the dispersion surface shown in figure 5.23. If the crystal is taken to be at the exact Bragg position A and B are the points corresponding to the waves in the first part of the specimen. Lines drawn through these points normal to the plane of the fault intersect the dispersion surface at A, B, C and D. These are the points that represent the waves in the second part of the crystal that can be matched at the fault

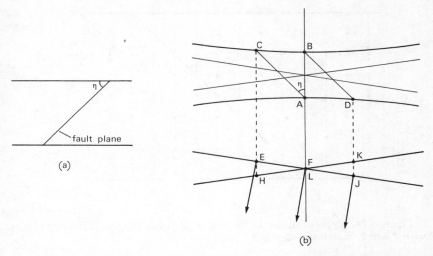

Figure 5.23. (a) Representation of a crystal containing a stacking fault; (b) the corresponding dispersion surface. A and B represent the waves above the fault, and A, B, C and D those below. At the exit surface these match to transmitted waves represented by E, F and J, and to diffracted waves represented by H, K and L

to those above. At the exit surface of the specimen the vectors of the waves change again because of the average potential (refractive index) of the crystal, and outside the crystal the waves are represented by the points E, F, H, J, K and L which lie on spheres of radius χ_0 centred upon O and G.

While the detailed form of the contrast can be found only by carrying out the matching process at the fault and finding the direct and diffracted wave intensities, many of the features of interest can be determined using only relatively simple arguments.

In the first place, since the components of the direct wave (points E, F and J) are travelling in slightly different directions, as also are the components of the diffracted wave (points H, L and K), fringes will be observed in the bright- and dark-field images. An exception to this is the case where the fault plane is not inclined to the surface of the sample, as then all these wave points are superimposed at one point. Another exception occurs when $g \cdot R$ is zero or an integer. When $g \cdot R$ is zero R is parallel to the reflecting planes, so that these are not disturbed by the fault, and hence there is no contrast. When $g \cdot R$ is an integer, the relative movement normal to the reflecting planes due to the fault is equal to a whole number of interplanar spacings. As a result, the reflecting planes match up across the fault and appear continuous, and again there is no contrast (figure 5.24).

The depth periodicity of the fringes in the image depends upon the extinction distance of the reflection being used, the deviation parameter

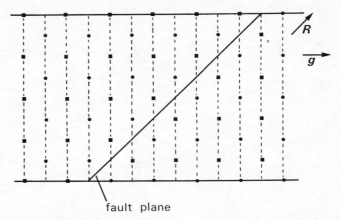

Figure 5.24. Representation of the case $g \cdot R = 1$, when the diffracting planes, shown dashed, match up across the fault

and the total specimen thickness. In the situation where the crystal is thin, so that absorption is relatively unimportant, all the waves A, B, C and D reach the exit surface of the specimen. Fringes occur in the bright-field image with spacings determined by the angle θ between EO and FO (and between FO and HO), and by the angle 2θ between EO and HO. Since EH is $2 \tan \eta / \xi_g$, 2θ is $2 \tan \eta / K\xi_g$ so that the possible depth periodicities are ξ_g and $\xi_g / 2$ in both bright- and dark-field images. The fringe pattern under these conditions is thus fairly complicated, consisting of strong fringes with a depth periodicity of ξ_g and weaker fringes halfway between them. Away from the reflecting position these fringe separations are reduced by $\sin \beta$.

Because absorption acts preferentially on the type 2 Bloch waves, only waves of type 1 are transmitted through a thick specimen with appreciable strength, and this leads to a simplification of the fringe pattern. Under these conditions, where the fault is close to the exit surface, only the type 1 wave (wave point A) reaches the upper side of the fault and so only the waves represented by the points A and C are present below the fault. Because the fault is close to the bottom of the crystal both of these waves reach the exit surface, where they give rise to the waves corresponding to the points E, F, H and L. As EF and HL are each $\tan \eta / \xi_g$ the depth periodicity of the fringes in bright and dark fields is ξ_g. These fringes are produced by interference between a type 1 and a type 2 wave: hence the bright- and dark-field images will be complementary, since this type of interference corresponds to the transfer of electrons back and forth between the transmitted and the diffracted directions.

On the other hand, where the fault is close to the entrance surface both of the Bloch waves initially present reach the fault, where they match onto waves represented by all the points A, B, C and D. Of these, the two type 2 waves (points B and C) are strongly absorbed so that only the other two waves (A and D) reach the exit surface, where they give rise to the waves, F, J, K and L outside the crystal. Again the depth periodicity is ξ_g but in this case, since the fringes are produced by interference between two type 1 waves, the images are not complementary but similar.

Near the centre of the fault the fringes tend to disappear almost completely. This is because only the type 1 wave (point A) reaches the fault, where it matches onto two waves, points A and C, in the lower part of the specimen. Of these, only the type 1 wave reaches the exit surface with appreciable strength, and since there is no other wave to interfere with no fringes occur.

To summarise, a fault in a thin crystal is expected to show a complicated pattern of fringes whose depth periodicity is $(\xi_g \sin \beta)/2$. As the crystal becomes thicker alternate fringes disappear, so that the fringe spacing becomes $\xi_g \sin \beta$ in depth, with low contrast at the centre of the image. The bright- and dark-field images are complementary where the fault is near the bottom of the crystal and similar where it is near the top, enabling the sense of the inclination of the fault to be found. By counting the fringes at the reflecting position the crystal thickness is obtained, and measuring the image width then enables the angle of inclination of the fault to be calculated. Hence, knowing the foil normal and a direction in the plane of the specimen, the fault plane itself can be determined.

Because of the effects of anomalous absorption the amplitudes associated with the points E, F, H, J, K and L are not constant but vary, as described above, with position across the image of the fault. Consequently, while it is possible to calculate these amplitudes and hence the fringe contrast, it is easier to use the column approximation described earlier. The waves present immediately below the fault are thought of as travelling together down a column through the crystal to the exit surface where they match onto a direct and a diffracted wave. The intensities of these waves vary across their wavefronts, corresponding exactly to the intensity variations obtained by combining the separate components outside the crystal as described above.

As a first step in calculating the contrast, suppose the wave corresponding to point A in figure 5.24 is incident upon the fault at depth z, at which point its amplitude is given by

$$[C_0^{(1)} \exp (2\pi i k_0^{(1)} z) + C_g^{(1)} \exp (2\pi i k_g^{(1)} z)] \exp (-\mu^{(1)} z) \qquad (5.43)$$

This is matched at the fault to a combination of the waves represented by the points A and C. Neglecting the changes in C_0 and C_g for the waves C, D due to small change in deviation parameter, we get

$$C_0^{(1)} \exp (2\pi i k_0^{(1)} z) \exp (-\mu^{(1)} z) = pC_0^{(1)} \exp (2\pi i k_0^{(1)} z) + qC_0^{(2)} \exp (2\pi i k_0^{(2)} z) \qquad (5.44)$$

Similarly, on matching the diffracted amplitudes we get

$$C_g^{(1)} \exp (2\pi i k_g^{(1)} z) \exp (-\mu^{(1)} z) = pC_g^{(1)} \exp (i\alpha) \exp (2\pi i k_g^{(1)} z)$$
$$+ qC_g^{(2)} \exp (i\alpha) \exp (2\pi i k_g^{(2)} z) \qquad (5.45)$$

These simultaneous equations can be solved for p and q, the amplitudes of the waves corresponding to points A and C. The other wave incident upon the fault can similarly be matched to a combination of waves (points B and D), and their amplitudes also found.

The waves in the second part of the crystal are now multiplied by absorption coefficients of the form $\exp\left[-\mu^{(i)}(t-z)\right]$, and by propagation factors of the type $\exp\left[2\pi i k^{(i)}(t-z)\right]$ (where $i = 1, 2$, as appropriate) to give the amplitude and phase of each wave at the bottom of the crystal. The transmitted amplitudes of these waves can then be added together to give the total transmitted amplitude, which is

$$\varphi_0(t) = \{\cos(\pi\Delta'kt) - i\cos\beta\,\sin(\pi\Delta'kt)$$
$$+ \tfrac{1}{2}\sin^2\beta\left[\exp(i\alpha) - 1\right]\cos(\pi\Delta'kt)$$
$$- \tfrac{1}{2}\sin^2\beta\left[\exp(i\alpha) - 1\right]\cos(2\pi\Delta'kt)\}\exp(-\pi t/\xi_0') \quad (5.46)$$

Δk is $|k^{(2)} - k^{(1)}|$, t_c is the distance of the fault below the centre of the specimen (that is, $t_c = z - t/2$) and $\Delta'k = \Delta k + i\sin\beta/\xi_g'$, so that $\Delta'k$ takes into account the differences in both the real and imaginary parts of the electron wave vectors. A similar expression is obtained for the diffracted amplitude

$$\varphi_g(t) = \{i\sin\beta\,\sin(\pi\Delta'kt) + \tfrac{1}{2}\sin\beta\left[1 - \exp(-i\alpha)\right][\cos\beta\,\cos(\pi\Delta'kt)]$$
$$- \tfrac{1}{2}\sin\beta\left[1 - \exp(-i\alpha)\right][\cos\beta\,\cos(2\pi\Delta'kt_c)]$$
$$- i\sin(2\pi\Delta'kt_c)]\}\exp(-\pi t/\xi_0') \quad (5.47)$$

A phase factor, $\exp\left[i\pi(\gamma^{(1)} + \gamma^{(2)})t\right]$, which does not affect the resultant intensities has been removed from each of these amplitudes.

We note that the average absorption, depending on the term in ξ_0', does not affect the form of the fringes but simply reduces the overall level of intensity. Where the fault intersects the surfaces of the specimen (that is, $t_c = \pm\tfrac{1}{2}t$) the amplitudes are those for a perfect crystal ($\alpha = 0$), so that the fault fringes match onto the intensity given by the adjacent perfect crystal. Also it can be seen that the bright-field image is symmetrical (since $\cos(2\pi\Delta'kt_c)$ has the same value when t_c is both $+$ and $-t/2$) whereas the dark-field image will not be, in agreement with the previous discussion where it was shown that the images of the fault were similar for the upper and complementary for the lower edges.

Some examples of the contrast calculated using this expression are shown in figure 5.25, where the phase angle of the fault has been taken to be either $+2\pi/3$ or $-2\pi/3$, the possible values for faults in the simple f.c.c. metals. These curves show the change in fringe periodicity from $\xi_g/2$ to ξ_g with increasing thickness, and the similarity and complementarity of the images at the upper and lower edges of the fault respectively. It can also be seen, and this is a general result, that the first fringe in the (symmetrical) bright-field image is bright when the fault angle α lies in the range 0 to π, and dark when it lies in the range 0 to $-\pi$ (or, equivalently, π to 2π). Thus the intensity of the first fringe gives information about the sign of α, and this information for several reflections together with a knowledge of the crystal structure of the material, and hence the likely fault vectors, may often enable the fault vector to be determined.

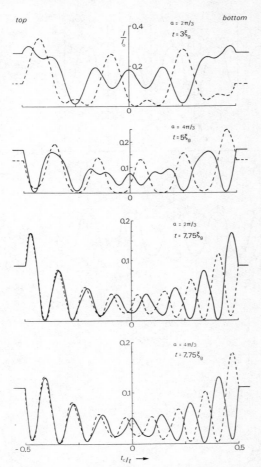

Figure 5.25. Calculated stacking fault contrast for a number of crystal thicknesses and fault angles of either $\pm 2\pi/3$: in all cases $\xi'_0 = 10\xi_g$, $\xi'_g = 15\xi_g$

Comparing the detailed form of the contrast with calculations can confirm the correctness of a vector found in this manner, or help in finding it in the case of a complex structure.

One special case that does not fit into the above categories is that of $\alpha = \pi$. This corresponds to the reflecting planes being displaced by half the interplanar spacing, as shown in figure 5.26a, and as a consequence the type 1 wave incident upon the fault can continue into the crystal below as a type 2 wave, without any type 1 wave being needed to complete the matching. Similarly, the type 2 wave propagates into the lower part of the crystal as a type 1 wave only. Because of this the bright- and dark-field images are both symmetrical and pseudo-complementary, even when absorption is allowed for. In thin crystals fringes of spacing $\xi_g/2$ in

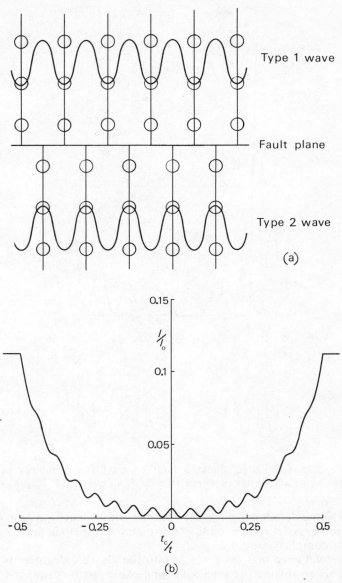

Figure 5.26. (a) The positions of the reflecting planes above and below a fault for which $\alpha = \pi$, $(\mathbf{g} \cdot \mathbf{R} = \frac{1}{2})$. The type 1 Bloch wave is able to pass directly into the crystal below the fault, where it is a type 2 wave: the type 2 wave similarly becomes a type 1 wave upon crossing the fault plane. (b) Calculated contrast for a fault of this type. The crystal thickness is $6\xi_g$ and $\xi'_g = 15\xi_g$

thickness are observed, but in sufficiently thick crystals the fringes disappear. The fault is still visible as a darker region with an intensity minimum at the centre of the image, as shown in the calculated profile in figure 5.26b. The centre is dark because in a thick crystal only the type 1 wave reaches the fault, where it becomes a type 2 wave in the lower part of the specimen. This wave is absorbed before it reaches the exit surface, and so the resultant intensity is low. At the upper edge of the fault both waves reach the fault, and so there is a type 1 wave below the fault that can travel to the exit surface and give some intensity. Similarly, at the lower edge only the type 1 wave reaches the fault where it is converted to a type 2 wave. Although strongly absorbed the distance to the exit surface is small, and so this wave emerges with appreciable strength, again giving intensity in the bright- and dark-field images.

Although there are many planar defects which, especially in non-stoichiometric structures, are more complicated than the simple stacking fault, their contrast is still found by applying the techniques used here. For computer calculations of fault contrast the problem is usually expressed in matrix form (Hirsch *et al.*, 1965, p. 223). In this rather more elegant description the initial direct and diffracted waves are written as a vector that is multiplied by suitable matrices that represent the process of transmission through a perfect crystal, matching at the fault, etc. This representation also enables the contrast from overlapping faults to be computed more straightforwardly. Examples of other boundaries that can be treated by this type of approach are twin boundaries and antiphase boundaries. A general treatment of the contrast at boundaries has been given by Gevers (1971).

5.4.3 Moiré Fringes

Moiré fringes are the fringes produced by superimposing two sets of parallel lines in the manner indicated in figure 5.27. In one case the sets of lines are parallel to each other, but the spacing in one set is slightly different from that in the other, while in the other the line spacings are the same but one set is rotated slightly relative to the other. Fringes are also produced by a combination of these two types of mismatch.

Moiré fringes can be formed in the electron microscope if two overlapping crystals have reflecting planes which, viewed edge-on, look like the sets of lines in figure 5.27. Thus the g-vectors of the planes are either parallel and almost equal, or else equal but slightly inclined to one another, and in addition fringes also occur if the g-vectors are both inclined and almost equal. These fringes have the same direction and spacing as the fringes which would be produced if the reflecting planes acted as absorbing strips. Their occurrence may be understood either by treating the crystals as thin, and assuming that the scattering is very weak, or more correctly, by considering them to be thick and treating the diffraction dynamically.

In cases where absorption is important, it can be shown that the dark fringes do in fact occur in the positions shown in figure 5.27, that is, where

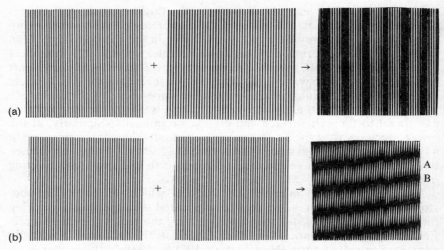

Figure 5.27. Formation of Moiré fringes by two sets of overlapping lines: (a) parallel but with different spacings; (b) the same spacing but rotated relative to each other

the sets of reflecting planes are out of register. The detailed treatment is similar to that employed in the case of a stacking fault, and the electrons are represented in the first crystal by the usual combination of two Bloch waves. These are matched at the interface to waves in crystal 2, the difference between this case and the stacking fault being that, in the case of the fault the reflecting planes on either side of the fault are out of register by a constant amount R, whereas in this case the mismatch varies periodically from exact coincidence at some points to a displacement of half the interplanar spacing at others. Thus the interface can be treated as a stacking fault with variable R, and the quantity $2\pi g \cdot R$ in the treatment of a fault can be replaced by $-2\pi \Delta g \cdot r$, which gives the local value of α at r in the interface (Gevers, 1962). The detailed calculation proceeds as in the case of a fault, but the positions of the fringes can be found without obtaining a complete solution if it is assumed that the crystals are sufficiently thick that only the type 1 wave in the upper crystal reaches the interface. Where the reflecting planes are in register, such as at A in figure 5.27, the type 1 wave is able to continue into the lower crystal unaltered since at these points the reflecting planes join up smoothly. Since the type 1 is well-transmitted, this wave reaches the exit surface and the image is locally light. On the other hand, at a point such as B, where the planes are exactly out of register, the wave in the upper crystal matches to a type 2 wave in the lower crystal (figure 5.26a). This wave, since it is strongly absorbed, does not reach the lower surface of the specimen so that in that region the image is dark. Intermediate degrees of mismatch at the interface give rise to intermediate amounts of intensity. Hence the dark fringes occur where the reflecting planes are out of register, as they do in the case of overlapping absorbing strips. This argument also shows that

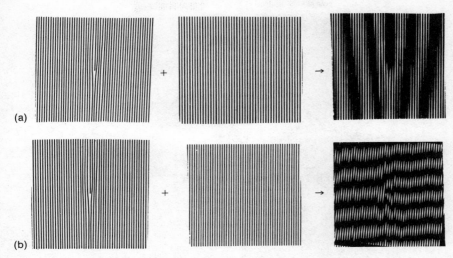

Figure 5.28. Moiré fringes produced by two overlapping lattices, one of which contains an edge dislocation: (a) the planes are parallel but have different spacings; (b) the spacings are the same but the planes are rotated relative to each other

the bright- and dark-field images will be similar in contrast. If the depth of the interface varies then the Moiré fringes are modulated by thickness fringes, while if Δg, the mismatch in reciprocal lattice vectors, does not lie in the interface the fringes are yet more complicated. Gevers (1971) includes Moiré fringes in his general treatment of the contrast at boundaries.

These arguments which give the fringe spacing and position also apply if one of the crystals contains an imperfection which distorts the reflecting planes in that crystal and hence alters the pattern of mismatch in the boundary. Thus, if one of the crystals contains a simple edge dislocation of the type shown in figure 5.28 then the Moiré fringes are correspondingly modified, as is also shown in the figure. These Moiré fringes look rather like lattice plane images, in which simple defects can also be seen. In this case, however, the apparent direction of the extra half-plane of the edge dislocation is different from the true direction. An example of Moiré fringes is shown in figure 5.29.

Moiré fringes provide a sensitive method for observing the mismatch between two lattices at an interface. They may occur, for example, at the surfaces of precipitates and at the boundary between a substrate and an epitaxially deposited layer. The mismatch at such an interface may also be accommodated by a suitable arrangement of dislocations lying in the interface. In principle, dislocations and Moiré fringes can be distinguished since the spacing and direction of the fringes alters with the reflection chosen, whereas the number of dislocations present and their directions are constant (although the number visible may vary with reflection). However, in cases where the mismatch is such that the fringe separation

Figure 5.29. Moiré fringes in stainless steel due to the mismatch between the lattice of the precipitates of $M_{23}C_6$ (where M is Fe or Cr) and the matrix of austenite. The mismatch is about 1.3 per cent. (Lewis and Hattersley; courtesy of *Acta metall.*)

is less than the width of the corresponding dislocation image the distinction between the two situations is difficult to make (Tholen, 1970).

5.5 CONTRAST FROM CONTINUOUS STRAIN FIELDS

Defects with continuous strain fields include dislocations, loops, voids and misfitting precipitates, and in each case the departure from perfection occurs throughout a region of the crystal surrounding the defect rather than at a single interface. The contrast due to a strain field of this type is found by solving the equations derived in section 5.3. Because of the forms of $R(z)$ that occur in practice it is not possible to find the wave amplitudes analytically, and they have to be calculated numerically. However, as in the case of stacking faults, useful information can be obtained without resorting to detailed computations.

According to the coupled equations of section 5.3 contrast arises because in a strain field the rate of transfer of electrons between the direct and diffracted waves is locally different. On the other hand, stacking fault contrast has been treated in terms of the transfer of electrons between one type of Bloch wave and the other. These two methods of calculating contrast are, of course, physically equivalent even if superficially they appear different. By treating a strain field as a very large number of small faults, in the manner indicated in figure 5.30, a dislocation or other defect

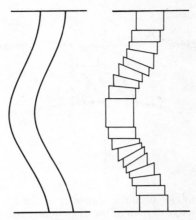

Figure 5.30. Sketch indicating how the displacement of the column on the left due to a defect can be represented by a sequence of stacking faults, as on the right

of this type may be treated using the approach adopted in section 5.4, when the contrast is identical to that obtained starting with the coupled equations. The two treatments can be shown to be formally equivalent. A slab of thickness dz can be approximated to a fault with phase angle $\alpha = 2\pi\{d[g \cdot R(z)]/dz\}\, dz$, and equations 5.44 and 5.45, together with the equivalent equations for the other Bloch wave, give, assuming α to be small (see Hirsch *et al.*, p. 238), and writing $\psi^{(i)}$ for $b^{(i)}(k,g)$

$$\frac{d\psi^{(1)}}{dz} = 2\pi i \frac{d}{dz}[g \cdot R(z)][\sin^2(\beta/2)\psi^{(1)} - \sin(\beta/2)\cos(\beta/2)\,e^{2\pi i \Delta kz}\psi^{(2)}]$$

(5.48)

$$\frac{d\psi^{(2)}}{dz} = 2\pi i \frac{d}{dz}[g \cdot R(z)][\cos^2(\beta/2)\psi^{(2)} - \sin(\beta/2)\cos(\beta/2)\,e^{-2\pi i \Delta kz}\psi^{(1)}]$$

(5.49)

These equations can be useful in understanding some aspects of image contrast, especially effects due to absorption. They show that if $R(z)$ varies slowly with z (for example far from a dislocation) the amount of scattering from one Bloch wave to the other is small, and hence the contrast is weak. On the other hand, if the strain field is large and localised, so that $dR(z)/dz$ is large, then a lot of scattering of the type shown in figure 5.31 will occur. A strain field of this type has short wavelength components that give scattering to points on the dispersion surface well away from those of the initial Bloch waves. A spread of points along the dispersion surface corresponds to a spread of directions of travel of the electrons, and thus to diffuse scattering characteristic of the strain field. The equivalent X-ray scattering (line broadening) was an early method of investigating defects in crystalline materials.

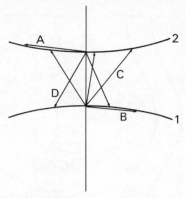

Figure 5.31. Strain fields produce scattering of the Bloch waves: this can be either to a point on the same branch of the dispersion surface, as at A and B, or to the other branch as at C and D: it is these latter transitions that give rise to contrast

5.5.1 Dislocations

The displacements due to a dislocation can be calculated (apart from the core) to various levels of sophistication, using elasticity theory. The contribution from the core is not normally thought to be important, but the atomic positions that have been calculated for various materials could be used to include the effect of this region also. If isotropic elasticity is assumed, the displacement R due to a screw dislocation is given by (Hull, 1965, Weertman and Weertman, 1964)

$$R = b\theta/2\pi \qquad\qquad (5.50)$$

θ is the angle measured in a plane normal to the dislocation line from an arbitrary origin, b is the Burgers vector, and it can be seen that the displacement is always parallel to b and to the dislocation line. Changing the origin from which θ is measured has no effect on the calculated contrast, since it alters all the displacements by a constant amount. The displacements around an edge dislocation are more complicated, but the main component at any point is again parallel to b. A dislocation of an intermediate type may be treated by resolving its Burgers vector into edge and screw components, and treating it as the superposition of two dislocations, one of screw and one of edge type. Consequently the displacements due to any dislocation are, to the first order, parallel to b. It follows that since there can be no contrast when $g \cdot R$ is everywhere zero, there will be little or no contrast if $g \cdot b$ is zero. This fact can be used to determine the Burgers vectors of the dislocations present in a crystal by taking a suitable set of micrographs of a thin area. If a dislocation is out of contrast in two micrographs, each taken under two-beam conditions, using different non-collinear reflection vectors, then b for that dislocation is the direction perpendicular to both of these vectors. Micrographs in which dislocations show no, or very little contrast, are shown in figure 5.32.

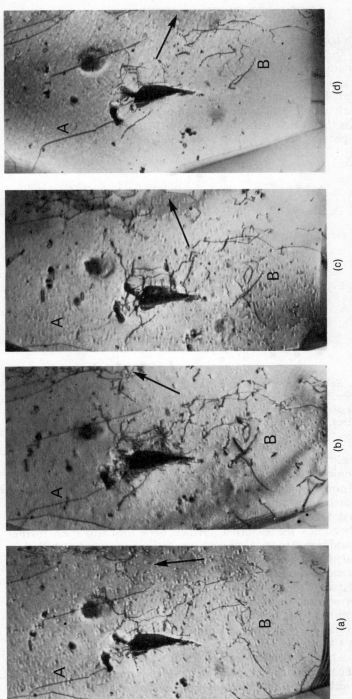

(a)　　(b)　　(c)　　(d)

Figure 5.32. A sequence of micrographs of stainless steel taken using the diffraction vectors indicated. The dislocation at A shows weak residual contrast in (c); the dislocation at B is out of contrast in (a) and shows a variety of contrast effects in the other micrographs. The dislocation at B also shows 'zig-zag' contrast due to dynamical effects. Note also the difference in strength of contrast and width of image of the dislocations at A and B in (b) and (a). The arrow on each micrograph represents the trace of the diffraction vector used in forming the image

Residual contrast arises because the displacements are not exactly parallel to *b*. In the case where a dislocation is wholly or partially edge-like in character, there is a component of the displacement that is normal to the slip plane (that is, the plane defined by *b* and the length of the dislocation). This displacement, illustrated schematically in figure 5.33, is relatively small but can give rise to residual weak contrast when $g \cdot b = 0$. This is the case for the dislocation at A in figure 5.32. Since this component is parallel to $b \times \hat{l}$ ($\hat{l}$ is a unit vector along the dislocation)

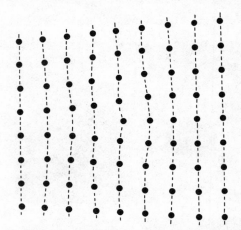

Figure 5.33. Schematic representation of the displacements in the vicinity of an edge dislocation. There are displacements normal to *b*, so the planes which are vertical are bent and can give residual contrast, even though $g \cdot b$ is zero when these planes reflect. (After Howie)

zero contrast only occurs when $g \cdot b$ and $g \cdot b \times \hat{l}$ are both zero simultaneously. Anisotropic elasticity of the crystal can also cause displacements giving residual contrast, as also can displacements due to surface relaxation. These latter arise because there will, in general, be stresses across any near plane to a defect within a large volume of material. Consequently, if such a plane becomes the external surface of a thin foil then further displacements additional to those given by equations such as 5.50 based on the simple theory will occur so as to make it stress-free. These displacements are most marked where a dislocation comes very close to a surface or meets it, and except in special cases they are not parallel to *b*. The effect of elastic anisotropy on the other hand does not depend upon the position of the dislocation, but will depend upon *b* and $\hat{l}$ (see Steeds, 1973). It can be small in a material such as aluminium or substantial as in the case of β-brass, which although cubic is sufficiently anisotropic that it will give strong contrast when $g \cdot b = 0$ (Humble, 1967). In such cases, comparison of the observed contrast and detailed calculations is needed to establish *b*.

The image width can be estimated by considering $g \cdot (dR/dz)$, which is the local change in *s* due to the strain field of the defect. For a screw dislocation lying in the plane of the specimen the value of $g \cdot (dR/dz)$ at the same depth as the defect and a distance x away is given by

$$g \cdot \frac{dR}{dz} = g \cdot \frac{b}{2\pi} \frac{d\theta}{dz} = g \cdot \frac{b}{2\pi x}$$

When $g \cdot b = 1$ this is $1/2\pi x$. A thick crystal reflects strongly over a range of $\pm 1/\xi_g$ in s, so that if it is assumed that a local change in s of $1/\xi_g$ or more will give appreciable contrast, the dislocation image will have a width of the order of d given by

$$1/\xi_g = 1/2\pi(d/2)$$

This gives a value for d of ξ_g/π. Detailed calculations confirm that this is a reasonable estimate, and show that the best resolution is usually obtained by using a low-order reflection for which ξ_g is small. High-order reflections, in general, have large extinction distances and tend to give diffuse images. A large extinction distance implies a small angular range of reflection and hence greater sensitivity to misorientations, so that the small orientation changes that arise a long way from a dislocation can affect the strength of such a reflection and give contrast. This is especially noticeable in the case of X-ray images (chapter 6): X-ray reflections have relatively very large extinction distances and hence great sensitivity to changes in orientation, and give very wide images of dislocations (typically several μm).

Simple arguments can also explain why dislocations appear as dark lines in both bright- and dark-field images. This is an effect of absorption and occurs for the same reason that the centre of a stacking fault is darker than adjacent areas of perfect crystal. The dislocation causes electrons to be transferred from one Bloch wave to the other, and since, because of absorption, the type 2 wave is generally the weaker of the two, it gains electrons from the type 1 wave. Below the defect the type 2 wave is again strongly absorbed, and in a thick crystal only the type 1 wave reaches the exit surface. Since it lost some electrons to the type 2 wave, it is weaker in the region of the dislocation than elsewhere, and so the dislocation appears as a dark line in both bright- and dark-field images. Where the dislocation approaches the upper surface both types of wave reach the defect, (or, at the lower surface, both types of wave leaving the defect reach the surface), so that both Bloch waves participate in causing contrast. As in the case of an inclined stacking fault the parts of the dislocation near the surface show oscillatory contrast due to interference between the two waves (figure 5.32).

If the strain field of the dislocation is represented as a set of faults in the manner of figure 5.30, the phase angles of the faults on one side of the dislocation will be opposite in sign to those on the other side, where the displacements are in the opposite sense. Consequently, a bright fringe on one side of the defect will be matched by a dark fringe on the other, giving the characteristic 'zig-zag' appearance to the image (as at B in figure 5.32).

It is also observed that except at the exact Bragg orientation, the image of a dislocation is usually displaced to one side of the dislocation line itself, and that the direction of the displacement changes with the sign of s. Figure 5.34 shows a micrograph of a dislocation crossing a bend contour, and the image on one side does not line up with that on the other side, where s has the opposite sign. Calculations show that the displacement is comparable with the image width, as is seen in the micrograph, and that the displacement is to the side that is closer to the reflecting orientation. This is reasonable, since on this side of the dislocation the scattering between the direct and diffracted waves is increased.

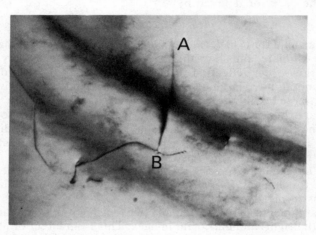

Figure 5.34. Micrograph of a dislocation (A–B) crossing a bend contour, showing the relative displacement of the images on the two sides

More detailed studies of dislocation contrast, essential in cases where the crystal is very anisotropic or where partial dislocations are not well separated, require exact solutions to the coupled equations. Since the contrast will vary along an inclined dislocation, the image intensity across the defect has to be found at a large number of points along its length. Even so, plots of the intensity across the dislocation do not always enable the contrast to be visualised clearly, especially where the contrast varies in an oscillatory manner. To facilitate comparison with experiment, methods of simulating a two-dimensional image have been devised, in which either the image is displayed on a cathode-ray tube controlled by the computer output, or the image is built up by overprinting selected symbols on a sheet of paper, darker areas being produced by greater degrees of overprinting. An image formed by this second method is shown in figure 5.35.

At first sight it might be thought that the time needed to compute such an image would be excessive, but methods exist, based on the matrix approach mentioned earlier, by which the calculation for two dimensions is effectively reduced to a calculation along a line across the dislocation (Head *et al.*, 1973).

Dislocations control the plastic properties of crystals, and much of the

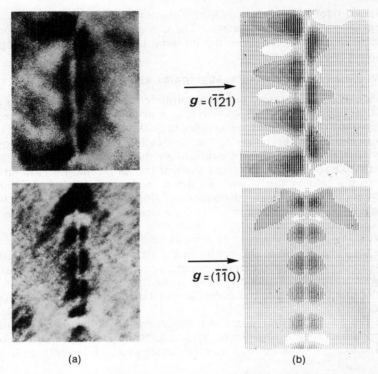

$g = (\bar{1}21)$

$g = (\bar{1}\bar{1}0)$

(a) (b)

Figure 5.35. (a) Two micrographs of a dislocation in β-brass and (b) the corresponding simulated images, showing the very good match obtainable. (Courtesy of A. K. Head)

effort expended in studying dislocations and their arrangements in deformed crystals has been aimed at understanding the details of the deformation process. The main problems in doing this are (a) the very small volume that can be studied in the microscope may not come from a typical region of the sample; and (b) because the microscope specimen is so thin the dislocations may re-arrange themselves or move out of the specimen upon removal of the long-range stresses present in the bulk material. Because of the large surface-to-volume ratio, experiments in which the specimen is deformed while under observation are less informative than they might be. These problems are reduced when a high-voltage microscope is used, since the thickness of the specimen can be greater. It has also been shown that in some cases neutron irradiation of the bulk sample produces radiation damage that pins the dislocations in the lattice, and reduces the amount of re-arrangement and loss that occurs when thinning the material (Mughrabi, 1968).

Dislocations also occur, especially in two-phase materials, because they are geometrically necessary in order that different parts of a crystal may fit together. Examples include the dislocations that occur at the interface when one material is deposited epitaxially on another and the lattices do not quite match, and the dislocations that occur at the surfaces of partially coherent precipitates. There is usually less tendency for such dislocations to be lost when the material is thinned for microscopy.

5.5.2 Small Loops, Vacancy Aggregates and Precipitates

Obviously a large loop, void or precipitate will be visible as such in the electron microscope. This section is concerned with situations where the dimensions of the defect are smaller than the extinction distance. Under these conditions, the image of the dislocation on one side of a loop will overlap that due to the dislocation on the other, and it is no longer permissible to treat the different parts of the loop separately. In all cases the strain field is fairly well localised, and occurs at a more or less well-defined depth. One consequence of this is that identical defects may appear to be different as a result of their being at different distances from the specimen surfaces.

We consider first the case of a small void or misfitting precipitate, and take the origin to be at the centre of the defect. This acts as a centre of pressure, giving displacements which, according to isotropic elasticity theory, are directed radially away from the centre of the defect. The displacement ϵ_r outside the defect at a distance r from the origin is given by $\epsilon_r = \delta/r^3$.

In the case of a precipitate the displacement within the precipitate is proportional to r. From figure 5.36, which shows these displacements schematically, it can be seen that the lattice planes intersecting the origin

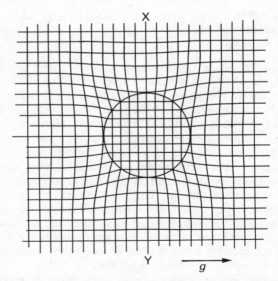

Figure 5.36. Schematic representation of the displacements around a coherent misfitting precipitate: the precipitate is in a state of uniform tension, whereas the matrix is in a state of shear

are not bent. Consequently there is a line of no contrast normal to g passing through the centre of the image. When the defect is near the centre of the foil the local intensity on either side of this line is, as in the case of a fault, below that of the rest of the specimen, due to the increased absorption arising from the transfer of electrons between the Bloch waves. Ashby and Brown (1963) investigated this problem and showed that the image size depends upon the parameter $\epsilon g r_0^3/\xi_g^2$. If the radius of the defect r_0 is known then ϵ, the degree of misfit can be found, and conversely, knowing ϵ enables r_0 to be determined. In addition, it is found that when the defect is near the surface the image becomes asymmetric and anomalously wide, and consists of a light region on one side of the no-contrast line, and a dark region on the other. This is again an effect of absorption, and as in the case of a fault, the bright- and dark-field images are similar when the defect is close to the entrance surface. Similarly, near the exit surface the images are complementary, the bright-field image having reversed its contrast, and the dark-field image being the same. From the dark-field image the sign of ϵ can be found, since if it is positive then the image goes from light to dark along the direction of g, whereas if it goes from dark to light then ϵ is negative: this is known as the Ashby–Brown rule. Some examples of the contrast due to small precipitates are shown in figure 5.37.

When the contrast expected from a misfitting inclusion is calculated

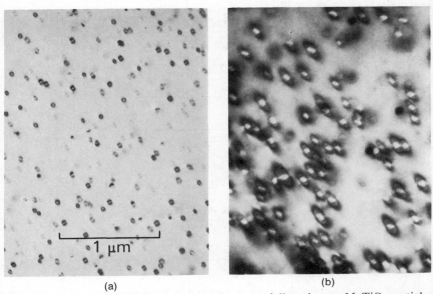

(a)   (b)

Figure 5.37. Strain-field contrast (a) from partially coherent $MgTiO_4$ particles in MgO; (b) rather larger precipitates of MgO in copper: even though in this case the particles are not spherical and are large enough to be resolved, the contrast is still recognisable as that of a misfitting inclusion. Note the line of no contrast normal to g. (Courtesy of M. H. Lewis)

using the displacements appropriate to an infinite crystal, it is found that the direction of the black–white contrast reverses for particles lying between $\xi_g/4$ and $3\xi_g/4$ from either surface, and also for the range $5\xi_g/4$ to $7\xi_g/4$, and so on. These reversals are not consistent with the Ashby–Brown rule; however, including the extra displacements that arise because of surface relaxation has the effect of reversing the direction of the contrast for the precipitates in the layer from $\xi_g/4$ to $3\xi_g/4$, so that all the inclusions within $5\xi_g/4$ of the surface have contrast whose sense is that predicted by the Ashby–Brown rule. Those in the layers from $5\xi_g/4$ to $7\xi_g/4$ from either surface do not fit in with the rule, but their contrast is weak (they are starting to make the transition from black–white to black–black contrast), and so the rule appears to be obeyed by all the most visible inclusions. The situation is illustrated schematically in figure 5.38, which is for the case of a precipitate that is too large for the hole it occupies in the matrix crystal.

Non-spherical particles and also anisotropic elasticity can give rise to images that are not symmetrical about the direction of $\mathbf{g}$. When the defects are small it may be difficult to distinguish between these two possibilities. Tilting the crystal well away from the Bragg orientation causes the scattering to become kinematical (see section 5.4.3), and the

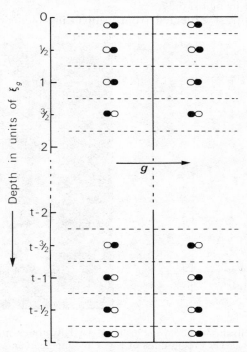

Figure 5.38. Sketch showing the direction of the black–white contrast from a coherent precipitate ($\epsilon > 0$) as a function of its distance from the specimen surface

contrast to arise mainly at the particle–matrix interface, so that in some cases its shape may be seen (Woolhouse and Brown, 1970). Usually, however, complete confirmation of the nature of the defect depends upon matching the observations with the contrast calculated using anisotropic elasticity theory.

Small loops, such as those formed as a result of radiation damage, give rise to contrast, producing displacements in the surrounding crystal that are due to all parts of the loop, as explained above. In the case of a prismatic loop, that can be formed by the aggregation of vacancies or interstitials, the displacements are not dissimilar to those given by a misfitting precipitate (Nabarro, 1967, § 2.1). This loop has the form shown in figure 5.39, with b normal to the plane of the loop. A shear loop, also

(a)   (b)

Figure 5.39. Schematic diagram showing the displacements at (a) a shear, and (b) a prismatic loop

shown in the figure, has a strain field, and hence contrast, with lower symmetry. In practice, loops may be mixed in character, with b inclined to the plane of the defect, and may also contain stacking faults that are visible when the loop is large.

However, it might be expected that prismatic loops at least would behave rather like small precipitates, and the type of loop could be found using the Ashby–Brown rule. In practice, contrast effects cause both directions of the black–white contrast to be visible, even in dark fields, when only one type of loop is present. Calculations show that, when using the displacements that include those due to surface relaxation the contrast reverses in the manner described earlier, so that loops in the range $\xi_g/4$ to $3\xi_g/4$ from the surface have contrast whose direction is opposite to that expected from the Ashby–Brown rule. As a result, it is less straightforward to determine the nature of the loops present in an irradiated or deformed material than to find the sign of the misfit for an inclusion. It is necessary to determine the depth of each defect within the foil and correlate this with the observed contrast. This can be done by taking a stereo pair of micrographs, from which the depth of each defect can be found using a parallax bar (Diepers and Diehl, 1966). It is best if the stereo pair is taken under kinematical conditions since this reduces the size of the image and thus makes the depth determination more precise, and an additional micrograph is taken under dynamical conditions to give the black–white contrast. When they lie close to the middle of a thick specimen, loops give black–black contrast in a similar manner to inclusions. Again, if the crystal is elastically very anisotropic, or the loops are not circular or have partial shear character, the resulting images may be asymmetrical, and final confirmation as to the precise nature of the defect will depend upon matching the observed contrast with that predicted by detailed calculations.

5.5.3 High-resolution Dark-field Microscopy

As a last example of the application of the ideas developed in the earlier part of this chapter we consider the weak-beam technique, which enables defects to be studied at high resolution. As pointed out previously, under normal two-beam conditions the width of the image of a dislocation or other defect with a continuous strain field is an appreciable fraction of an extinction distance ξ_g. If, however, the crystal is tilted a long way from the exact Bragg orientation the effective extinction distance is $\xi_g/(1 + w^2)^{\frac{1}{2}}$, ($\sim 1/s$) and the images produced under these conditions might be ex- to be very much narrower.

This is indeed the case, but the defects are then rather difficult to observe. The local deviation parameter $s + g \cdot (\mathrm{d}R/\mathrm{d}z)$, which appears in the coupled equations is very large for most of the specimen, and the diffracted intensity is correspondingly exceedingly weak. The only part of the specimen to diffract with appreciable intensity is that close to the defect for which

$$s + g \cdot \frac{\mathrm{d}R}{\mathrm{d}z} \sim 0 \tag{5.51}$$

However, when s is large the volume of material in which this condition is satisfied is so small that the depletion of the transmitted beam intensity is negligible, and so the defect is not visible in the bright-field image. On the other hand, most of the dark-field image has very low intensity because of the large value of s, and therefore the weak diffraction from the defect does show up, as a lighter region against the dark background. Using the value of $g \cdot (\mathrm{d}R/\mathrm{d}z)$ found previously for a screw dislocation, we find that the image width defined by equation 5.51 is approximately $1/\pi s$. Since $1/s$ may be made ten times smaller than ξ_g the image of the dislocation can be an order of magnitude narrower. The larger the value of s, the smaller the intensity, and the longer the exposure needed to record the micrograph, so that there is a limit to the technique set by the stability of the instrument. An example of a weak-beam image is shown in figure 5.40.

5.6 GUIDE TO FURTHER READING

There are several substantial books concerned almost entirely with electron microscope image contrast, and which therefore treat the material encompassed by this chapter in considerably greater detail. These

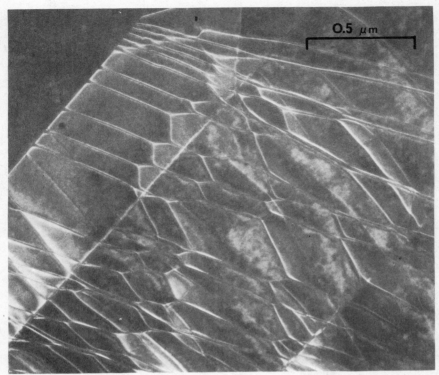

Figure 5.40. High-resolution dark-field image of dislocations in TaS₂. The deviation parameter, and hence the image width, varies across the field of view. (Courtesy of W. M. Stobbs)

include Hirsch *et al.* (1965), Amelinckx *et al.* (1970), and Valdrè and Zichichi (1972); all of these have been referred to on specific points in the text. The first two are concerned mainly with contrast from crystalline specimens, while the third includes a treatment of phase contrast. The book edited by Kay (1965), has a number of chapters concerned with the staining of biological materials in order to produce contrast.

REFERENCES

Amelinckx, S., Gevers, R., Remaut, G., and van Landuyt, J., (eds) (1970), *Modern Diffraction and Imaging Techniques in Material Science*, North-Holland, Amsterdam

Ashby, M. F. and Brown, L. M., (1963), *Phil. Mag.*, **8**, 1083, 1649

De Rosier, D. J., and Klug, A., (1968), *Nature*, **217**, 130

Diepers, H., and Diehl, J., (1966), *Phys. Stat. Sol.*, **16**, K109

Gevers, R., (1962), *Phil. Mag.*, **7**, 1681

Gevers, R., (1971), *Electron Microscopy in Material Science* (ed. U. Valdrè), Academic Press, New York, 302

Head, A. K., (1969), *Aust. J. Phys.*, **22**, 43, 345

Head, A. K., Humble, P., Clarebrough, L. M., Morton, A. J., and Forwood, C. T., (1973), *Computed Electron Micrographs and Defect Identification*, North-Holland, Amsterdam

Hirsch, P. B., Howie, A., Nicholson, R. B., Pashley, D. W., and Whelan, M. J., (1965), *Electron Microscopy of Thin Crystals*, Butterworths, London

Howie, A. and Sworn, C. H., (1970), *Phil. Mag.*, **22**, 861

Hull, D., (1965), *Introduction to Dislocations*, Pergamon, Oxford

Humble, P., (1967), *Solid St. Phys.*, **21**, 733

International Tables for X-ray Crystallography, (1962), (ed. Lonsdale), Kynoch, Birmingham, vol. 3, tables 3.3.3 A(1), A(2)

Kay, Desmond, (ed.) (1965), *Techniques for Electron Microscopy*, Blackwell Scientific, Oxford

Misell, D. E., and Childs, P. A., (1972), *J. Phys. D.*, **5**, 1760

Misell, D. E. and Childs, P. A., (1973), *J. Phys. D.*, **6**, 1653

Mughrabi, H., (1968), *Phil. Mag.*, **18**, 1211

Nabarro, F. R. N., (1967), *Theory of Crystal Dislocations*, Oxford University Press

Steeds, J. W., (1973), *Anisotropic Elasticity Theory of Dislocations*, Clarendon, Oxford

Tholen, A. R., (1970), *Solid St. Phys.* (a), **2**, 537

Thon, F., (1971), *Electron Microscopy in Material Science* (ed. U. Valdrè), Academic Press, New York, 568

Valdrè, U., and Zichichi, A., (eds) (1971), *Electron Microscopy in Material Science*, Academic Press, New York

Weertman, J. and Weertman, J. R., (1964), *Elementary Dislocation Theory*, Collier-Macmillan, London

Woolhouse, G. R., and Brown, L. M., (1970), *J. Inst. Metals*, **98**, 106

Yoshioka, H., (1957), *J. phys. Soc. Japan*, **12**, 618

6

X-ray Topography

We have seen how the transmission electron microscope permits one to see most kinds of defects in solids in tremendous detail and it is certainly the most widely used of the instruments described in this book. Even so it has its limitations, which mainly arise from the facts that the specimens have to be very thin and that the electron microscope does not work at all well below about 1000× magnification.

Thin foils sometimes behave differently from bulk matter. For example, dislocations easily slip out of a thin foil in a soft material, leaving behind a very unrepresentative structure; the type of magnetic domain boundary seen in thin foils is interesting in itself but quite different from those existing in larger specimens, and phase transformations carried out in foils may have special characteristics due to the high surface/volume ratio of the specimens. Another disadvantage of the small specimen dimensions is that the amount of material actually sampled is minute. A typical deformation experiment consists of deforming a specimen in a mechanical testing machine, sectioning it on various planes and then studying the dislocation structure in the electron microscope; a very thorough and skilful experimentalist might succeed in seeing through about 0.001 per cent of the total specimen volume.

It probably would be possible to design electron microscopes to work at low magnifications if necessary, but the results would be of little use. The image width of a dislocation (section 5.5) is about 5 nm, so dislocations cannot be resolved at low magnifications. It follows that the electron microscope is not suitable for the study of large-scale phenomena (such as the change of orientation of an imperfect crystal over a centimetre or so) or of rather perfect crystals, such as the semiconductor and laser crystals used in industry or metal crystals grown by special techniques. The lower limit of dislocation density which can reasonably be studied in the transmission electron microscope is about 10^4 mm^{-2} whereas it is now possible to grow large crystals of some materials which contain no dislocations at all. The development and control of such methods requires techniques which will reveal the degree of perfection of large-scale specimens (with dimensions of centimetres rather than microns), at quite low magnifications.

The methods of X-ray topography provide just these techniques. The higher penetrative power of X-rays, resulting from the lower scattering factors of atoms for X-rays relative to electrons, means that quite thick samples in the range of 10 μm to about 10 mm can be studied. Dislocations, stacking faults and minute misorientations are revealed by diffraction, and the images of defects are wide enough to be easily visible. Only crystalline specimens can be studied, and X-ray methods cannot be used at high magnification because their resolution is limited to about 1 μm. Likewise, they cannot be used for studying defects in work-hardened material, but, fortunately, the upper limit of dislocation density for X-ray topographic methods coincides (at about 10^4 mm^{-2}) with the lower limit for electron microscopy.

6.1 EXPERIMENTAL METHODS IN X-RAY TOPOGRAPHY

Images can be formed by X-ray diffraction by a method very similar to that of dark-field electron microscopy. A specimen is illuminated with a wide X-ray beam (or scanned through a narrow beam) and arranged so that a diffracted beam is reflected off one set of crystal planes. A photographic plate is used to map the intensity of the diffracted beam across the surface of the crystal. Defects within the crystal can locally alter the orientation and spacing of the reflecting planes, thus causing a local change in the diffracted intensity, and an image of the defect appears on the photographic plate. As in electron microscopy, images are caused by the strain fields of defects and it is necessary to consider the interaction between the strain fields and the X-ray wave fields in the crystal in order to understand the contrast of images.

The theory of diffraction from perfect and imperfect crystals is treated in section 6.2, but it is useful to have a qualitative idea of the basic contrast effects before describing the experimental arrangements. Three types of contrast can conveniently be distinguished as discussed below.

Orientation Contrast

Imagine two perfect regions of crystal that are misorientated with respect to one another, such as two sub-grains. If they are illuminated by a monochromatic beam (which is well collimated so that its divergence is less than the misorientation), one region may be set at the Bragg angle and give rise to a diffracted beam, while the other may be right outside the angular range of reflection and give a negligible diffracted beam. There will, therefore, be strong contrast between the regions. Alternatively, if continuous radiation is used both regions will diffract but the two diffracted beams will emerge at different angles (because the two regions will have selected different wavelengths) and again the regions can be distinguished on the photographic plate.

Direct Contrast

Under normal experimental conditions the diffracting power of imperfect crystals is much higher than that of perfect crystals (as explained in sections 6.2.2 and 6.2.3). If a specimen consists of a perfect crystal containing imperfect regions such as dislocations or precipitates then the photographic plate will record higher diffracted intensities from these regions and contrast will be achieved. Individual dislocations can easily be observed by this contrast mechanism.

Direct contrast is often called 'extinction contrast' from the strong attenuation of a beam set exactly on the Bragg angle in a perfect crystal—the 'primary extinction' effect. However, this label properly applies only to the Bragg or reflection case in which the incident beam is reduced to negligible intensity after traversing a few microns of crystal; in transmission the same attenuation might require hundreds or thousands of microns, and it is not called primary extinction. Unfortunately, contrast effects on transmission topographs are often loosely called extinction contrast. We consider that it is preferable to leave the classical terms properly defined and to use the terms 'direct contrast' and 'direct images' for effects that arise because imperfect regions directly reflect a higher fraction of the incident beam than do perfect regions.

Dynamical and Intermediary Contrast

When relatively thick, nearly perfect crystals are used in transmission a number of complicated effects occur, due to interference between Bloch waves, as in electron microscopy. These may have the reverse effect to direct contrast, for example relatively imperfect regions can cause a decrease in diffracted intensity. We shall postpone discussion of these effects until section 6.2, since they are governed by the type of specimen rather than by the experimental conditions: any apparatus that will produce images by direct contrast in transmission will produce intermediary or dynamical images if thicker specimens are used.

Many different experimental arrangements are possible for obtaining X-ray diffraction images and they all come under the general title of 'X-ray topography'. One may use point or line X-ray sources, continuous or monochromatic radiation, reflection or transmission with moving or stationary specimens. We shall classify them on the basis of the type of incident radiation (the review by A. R. Lang, 1970, should be consulted for more details of the experimental arrangements).

6.1.1 Continuous Radiation Methods

The experimental arrangements for reflection and transmission are shown in figure 6.1. Most X-ray topographic methods are known after their inventors and in this case the reflection arrangement is due to Schultz and the transmission method to Guinier and Tennevin. The X-rays sample about a 5 μm depth of the crystal in reflection (no more, because of the strong primary extinction) but crystals up to about 1 mm thick may be used in transmission.

The characteristics of these methods depend on the distance between the specimen and the film. If the film is placed as close as possible to the specimen (without being in the direct beam) direct contrast only will be

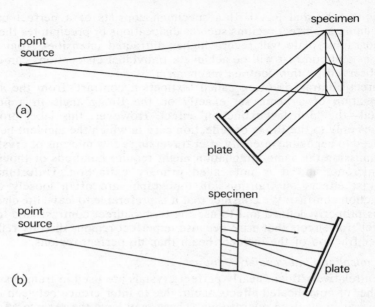

Figure 6.1. Continuous radiation methods: (a) the Schultz method for reflection, (b) the Guinier–Tennevin method for transmission

observed, but as the film is moved away from the specimen the spatial separation of images from misorientated regions increases and orientation contrast is observed. If the distance from source to specimen about equals that from specimen to film, good sensitivity to both types of contrast is found. With care, it is possible to image individual dislocations (Fujiwara *et al.*, 1964; Fiermans, 1964). Better methods are available for studying dislocations however, and this method is usually employed for a quick, simple investigation of defects such as sub-grains, or lineage structures found in melt-grown crystals. A Schultz photograph of a crystal containing sub-grains and lineage structures is shown in figure 6.2.

The sensitivity of either the reflection or the transmission methods to orientation changes is easily found. If the misorientation of two sub-grains is $\Delta\theta$ about an axis normal to the plane containing the incident and diffracted beams (the so-called 'incidence plane'), then the angle between diffracted beams from either grain is $2\Delta\theta$. If s is the specimen-to-film distance then the displacement of images on the film will be $2s\Delta\theta$. If x is the resolution of the film then the angular resolution of the system is

$$\Delta\theta = x/2s \qquad (6.1)$$

Putting typical figures of $s = 500$ mm, $x = 100$ μm for ordinary X-ray film, $\Delta\theta = 10^{-4}$ radians or about 20 seconds of arc. This can easily be improved by increasing s, decreasing x (the best plates can resolve less than 1 μm) or by utilising the original idea of Guinier and Tennevin shown in figure 6.3, in which the dimensions of the system are chosen to give a focusing

Figure 6.2. A Schultz photograph of a niobium crystal containing sub-grains and lineage structures. The field of view is about 4×10 mm

action which increases the sensitivity and ease of measurement of misorientations. The image of a sub-grain would be a line perpendicular to the plane of figure 6.3.

A point source of X-rays is essential for these topographic methods. An extension of the source in the direction perpendicular to the incidence plane, will blur the image of the sub-grain in the same direction. The blurring d caused by the finite source height in any topographic method is given by

$$d = \frac{hb}{a + b(1 - h/c)} \tag{6.2}$$

where h is the source height, a the source-to-specimen distance, b the specimen-to-film distance and c the height of the specimen. h/c is almost always negligible, so the equation reduces to

$$d = hb/(a + b) \tag{6.3}$$

If the specimen is about equidistant from the source and the film, $a = b$ and the resolution on the plate will be about half the diameter of the source. This is about 1 mm for a conventional X-ray tube, a few tens of microns for a microfocus tube, or less than a micron for sources formed by magnetic lens focusing of electron beams (as in the electron-probe

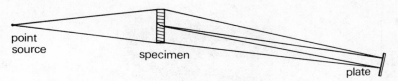

Figure 6.3. The Guinier–Tennevin focusing method for increasing the sensitivity of measurement of small misorientations

instruments described in chapters 2 and 3). Clearly this spatial resolution will directly affect the angular resolution figures discussed in the last paragraph, and unless a very fine source is available there is no point in using the highest quality films or plates. The source must also be narrow in the incidence plane itself for continuous radiation methods, but this is not essential for some methods using monochromatic radiation.

Continuous radiation methods can be used rather effectively with a special type of X-ray source available in a few laboratories in the world, that is, the synchrotron radiation emitted from electrons circulating in a synchrotron or storage ring. This is typically continuous radiation of wavelength $\geqslant 0.05$ nm, very small apparent source size (comparable with that in an electron-probe instrument) and colossal intensity—about 1000 times greater than that from the usual 2 kW sealed-tube generator. These continuous radiation methods have the advantage of simplicity of apparatus and manipulation; they also produce images from several different reflecting planes simultaneously, which speeds up quantitative analysis of the misorientations or of the strain fields causing direct contrast. However they do not usually give the highest resolution for images of dislocations and other defects and the details of the images can be difficult to interpret since the wavelength varies across the image.

6.1.2 Parallel Monochromatic Beam Methods

The basic arrangements for reflection and transmission are shown in figure 6.4. The reflection technique is called the 'Berg–Barrett' method

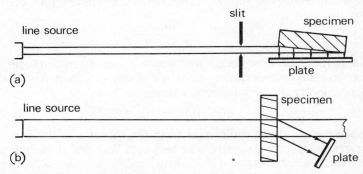

Figure 6.4. Parallel monochromatic beam methods: (a) the Berg–Barrett method for reflection, (b) the Barth–Hosemann method for transmission. Many minor modifications can be made to the basic geometry of these arrangements

after its invention by Berg and development by Barrett. Newkirk (1958, 1959) made further improvements and gave an excellent account of the experimental conditions required for best resolution.

The source is a line focus X-ray tube, the line (usually) being in the incidence plane. The incident beam is very divergent in the incidence plane but since the reflecting planes are set perpendicular to this plane at the Bragg angle for the K_α line, beams arriving at other angles cause little diffraction (since the K_α line is so much more intense than the other components of the spectrum). The line focus must again be very narrow in the direction perpendicular to the incidence plane, since a point on the specimen will diffract rays from an arc on the source as illustrated in figure 6.5. Equations 6.2 and 6.3 hold for the loss of resolution, and

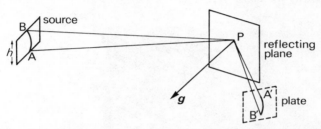

Figure 6.5. The effect of finite source height. A point P on the specimen will reflect rays from the arc AB on the source (since they all make the same angle with the reflecting planes) and will be blurred into an arc A'B' on the photographic plate

because the film is placed very close to the specimen $(a \gg b)$ in the Berg–Barrett method we may further simplify the formula to

$$d \approx hb/a \qquad (6.4)$$

h is typically 100 μm so b/a must be $\leqslant 0.01$ if the resolution of the best plates is to be fully exploited. The film is placed as close as possible to the specimen and a Bragg reflection chosen so that the incident beam makes a low angle with the surface, using slits to prevent the beam hitting the film (see figure 6.4). The film obviously cannot be placed perpendicular to the diffracted beam unless both the Bragg angle is near 45° and the reflecting plane is at about 45° to the crystal surface, so a relatively thin emulsion must be used, say 10 μm thick. If a 100 μm thick emulsion were used the diffracted beam would have to strike it at 90° ± 0.5° to avoid loss of resolution due to the inclination of the beam in the emulsion.

For the very highest resolution only the $K_{\alpha 1}$ line of the $K_{\alpha 1 \alpha 2}$ doublet should be used. If a suitable crystal monochromator is not available (section 6.1.4) the distance from the source to the specimen can be greatly increased; a value of $a = 10$ m might be needed and it is usually more convenient to use the slit-collimated method described in the next section. If this apparatus is not available it is usually possible to use the $K_{\beta 1}$ characteristic line, since for most X-ray target materials the other K_β lines are very feeble.

The Berg–Barrett method is one of the most commonly used methods of X-ray topography because of the relative simplicity of the apparatus, the high resolution of about 1 μm, and because it is very informative. Direct contrast is observed and it is possible with care to see individual dislocations. Sub-grains are seen by orientation contrast if their misorientation is large, by displacement contrast at the sub-grain boundaries if their misorientation is small (a displacement of different sub-grain images arises because their images are formed by incident beam components at slightly different angles) and by the direct image of the sub-grain boundary. Other surface features can be seen, such as a region of different composition and hence lattice parameter. A Berg–Barrett photograph of the same crystal as was used for the Schultz method (figure 6.2) is shown in figure 6.6.

Figure 6.6. A Berg–Barrett photograph of the same niobium crystal that was used for figure 6.2

In the corresponding transmission method (Barth and Hosemann, 1958, figure 6.4b) exactly the same considerations apply if high resolution is needed. However, it is not possible to place the film very close to the specimen in transmission because of the overlap of the transmitted and diffracted beams. The images obtained are similar to those taken with the Lang method described in the next section but the background intensity is higher and the resolution poorer; on the other hand, the apparatus is less expensive and the exposure times are shorter.

6.1.3 Collimated Monochromatic Radiation Methods

These are the most versatile, informative and widely-used of all X-ray topographic methods. The arrangements for reflection and transmission are shown in figure 6.7; the reflection method is a high-resolution variation of the Berg–Barrett method, first used by Merlini and Guinier (1957), and the very important transmission arrangement is known as the Lang method (Lang 1959; its inventor modestly calls it the 'projection topograph' method).

A point source of X-rays is used. The vertical height of the source will affect the resolution as before (equation 6.4). The horizontal divergence of the beam (that is, in the incidence plane) is governed by the horizontal width of the source, the width of the collimating slit S_1 and the separation of the source and slit; these parameters are chosen so that the divergence is less than the angular difference between $K_{\alpha 1}$ and $K_{\alpha 2}$ reflections. From the Bragg law

$$\lambda = 2d \sin \theta \tag{6.5}$$

we obtain by differentiation

$$d\theta = \frac{d\lambda}{\lambda} \tan \theta \tag{6.6}$$

which gives the maximum beam divergence. For example, for a reflection off the {110} planes of iron, using MoK_α radiation, the allowed divergence works out at about 10^{-3} radians or 220 seconds of arc.

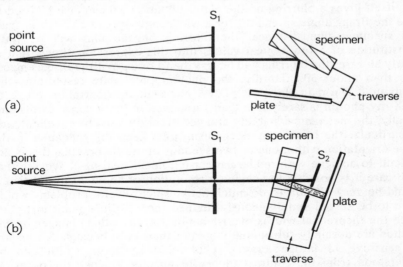

Figure 6.7. Collimated monochromatic radiation methods: (a) the Merlini–Guinier method for reflection, (b) the Lang method for transmission

A further slit S_2 is placed between the specimen and film to cut off the transmitted beam. The film is placed perpendicular to the diffracted beam so that thick emulsions can be used without loss of resolution. A photograph may be taken with all components stationary; this is called a 'section topograph' and, as seen by reference to figure 6.7b, gives an image of a slice *across* the crystal, enabling the depth of features to be determined with an accuracy that improves as the width of slit S_2 is decreased (S_2 might be made ~ 10 μm for a section topograph). The more usual procedure is to keep everything stationary except the specimen and film which, rigidly linked together, are scanned through the X-ray beam to cover the required area of the crystal. In the scanning method, the slit S_2 need only be narrow enough to stop the direct beam and other general scattering, say 500 μm. Exposure times are of the order of 10–25 hours per millimetre of scan with a conventional X-ray set and about ten times shorter with a high-power rotating anode set. These exposures are about ten times longer than those for the stationary crystal methods, since only the $K_{\alpha 1}$ components of the incident beams are used in order to obtain the best resolution. An example of a Lang topograph is shown in figure 6.8, in which individual dislocations are easily seen.

Since the Lang method provides the highest spatial resolution of the X-ray topographic methods it is worth considering in more detail the parameters affecting resolution. First, the vertical height of the source will limit resolution as already discussed, but since this may be controlled by means of the source–specimen distance (equation 6.4) it is not a limiting factor; the specimen–film distance is usually about 10 mm so that 1 μm resolution is obtained with a source–specimen distance of 1 m if the source itself is 100 μm high. Secondly, the wavelength spread in the $K_{\alpha 1}$ line itself gives a blurring on the plate. Equation 6.6 can also be used to give the Bragg angle spread $d\theta$ for a wavelength spread $d\lambda$; if b is again the specimen–film distance, the blurring on the film will be $b\,d\theta$. Substitution of a few typical values into this expression shows that in nearly all cases the blurring will be less than 1 or 2 μm if the Bragg angle is less than about 30°. Thirdly, the photographic plate resolution—this is about 0.25 μm in the best plates, although the ionisation range of an X-ray photon is rarely less than 1 μm even in these dense emulsions. Finally, the mechanical stability and construction must be excellent, and in particular the traversing mechanism must keep the specimen in the same orientation within one or two seconds of arc (in practice this is not difficult to achieve with modern precision linear bearings). We see that with care it is possible to achieve a resolution of about 1 μm but that it would be very difficult to do much better since several problems would have to be solved simultaneously; indeed there is little point in trying, since the theoretical widths of dislocation (or any other) images in this method are normally substantially greater than 1 μm because X-rays are so sensitive to lattice strains. This relatively large width can be understood from the contrast theory of chapter 5, since the order-of-magnitude arguments used in the electron case also apply to X-ray topography, in particular when dynamical images are present, and the

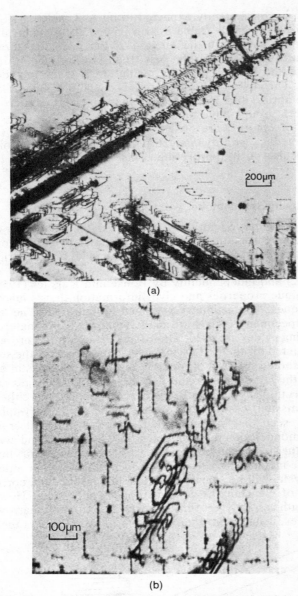

(a)

(b)

Figure 6.8. Lang topographs of dislocations in a 100 μm thick slice of silicon. The plane of the slice is (111) and the reflecting plane is one of the other {111} planes. MoK$_\alpha$ radiation, $\mu t = 0.15$. Dislocation pile-ups and sources are visible in (a) and the contrast is mainly 'direct' (extra intensity from distorted regions) though the oscillatory contrast of some of the straight dislocations seen at higher magnification in (b) indicates a small 'intermediary' character in the image. (Courtesy of J. Miltat)

image width is consequently expected to be $\xi_g/3$. Since ξ_g for X-rays is typically tens of microns (table 6.2 on page 232), dynamical images of dislocations are usually in the range 5–20 μm wide; the direct images, although narrower, are rarely less than 1 μm wide. An excellent discussion of the factors affecting resolution is given in the review by Lang (1970). The cost of a Lang camera is roughly comparable to that of an elaborate optical microscope (rather than an electron microscope) and many research groups have built their own.

All types of contrast are observed in Lang topographs: orientation, direct, intermediary and dynamical, depending on the type of specimen and in particular on its absorption. This will be discussed in detail in section 6.2.

6.1.4 Double-crystal Methods

Since the rocking curve width of a good crystal is only a few seconds of arc, highly monochromatic beams can be obtained by diffraction and used as incident beams for X-ray topography. We shall first describe a rather special case, the so-called 'parallel' setting developed by Bond and Andrus (1952) and independently by Bonse and Kappler (1958), since it has some unique properties and gives information that is inaccessible to other techniques. The geometry is shown in figure 6.9. The 'monochromator' and specimen crystals are made from the same material and the same reflecting planes are used. The Bragg angles are therefore equal and a beam diffracted off the first crystal will also diffract off the second. The first crystal should be perfect; defects in the second crystal cause local rotations of the lattice planes and will therefore give images on the film. The first reflection should be a Bragg case (reflection) but either Bragg or Laue (transmission) cases may be used at the second crystal.

It is easily seen that in this particular method the first crystal is not acting as a monochromator, since any wavelength spread will only be limited by the restricted range of incident angles. For high spatial resolution the incident radiation must be slit-collimated or a point on the specimen crystal will be smeared out to a line on the film, corresponding to the range of wavelengths getting through the system. However, any one wavelength can only get through over an extremely narrow range of incident angles, usually less than a second of arc. There is therefore, no

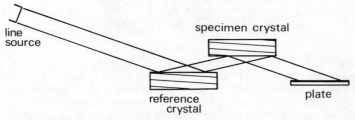

Figure 6.9. The double-crystal parallel setting. Both crystals are of the same material and the same type of diffracting plane is used in each case

strong direct image as in the Lang method, but the orientation contrast is exceptionally high. It is possible to detect orientation changes of fractions of a second of arc (corresponding to the strain field of a dislocation at 50 μm range) or changes in interplanar spacing of a few parts in 10^8 (Hart, 1968). Naturally, the setting of the second crystal must be as precise as the orientation sensitivity: it corresponds to a movement of about 1 μm at the end of a lever 1 m long.

The parallel setting is used when the highest strain sensitivity is wanted. Another use for multiple-crystal techniques is to provide a monochromatic and/or parallel incident beam for more conventional topography, as an alternative to slit collimation. If such beams can be produced then the parallel beam methods (section 6.1.2) can produce good results without the complication of scanning the specimen. The methods are useful for the same sort of applications as the Lang method (study of dislocations, stacking faults, etc.) and the contrast is broadly similar. The apparatus is slightly less complicated and exposure times usually less. However, the details of the image contrast are not so easily calculated because of the rather awkward relationship between the wavelength and the divergence in many double- or multiple-crystal arrangements, and the Lang method is usually preferable for detailed work.

One of these methods does, however, give almost ideal conditions for calculating the image contrast, that is, a divergent but highly monochromatic incident beam. This is the OMD (oscillating, monochromatic, divergent) method of Kohra and Takano (1968) illustrated in figure 6.10, in which a point source of X-rays is used with a curved crystal monochromator. The $K_{\alpha 1}$ and $K_{\alpha 2}$ components are spatially dispersed and $K_{\alpha 1}$ is selected with a slit at the focus of the monochromator. The specimen is oscillated to bring all parts into the Bragg position. High-quality images are obtained and the method has the great advantage that it produces equally good results on crystals which are elastically curved, or which contain a number of sub-grains; one frustrating feature of the original Lang method is that only one part of such crystals will be at

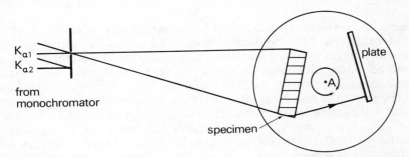

Figure 6.10. The OMD method. A curved monochromator (not shown) is used to separate $K_{\alpha 1}$ and $K_{\alpha 2}$. The crystal and photographic plate are oscillated about a common axis, to bring all parts of the crystal into the Bragg position. A is the axis of rotation, which may be anywhere

the correct setting and great care must be used to avoid elastic bending (a radius of curvature of about 25 m can easily be detected).

There are many other multiple-crystal methods that we shall not discuss since they are rather specialised and have been reviewed by Kohra *et al.* (1970) and Nakayama *et al.* (1973). It is worth pointing out that interferometric techniques using Moiré fringes are applicable to X-ray topography as well as to electron microscopy and of course the same theory applies (section 5.5). X-ray Moiré methods are capable of revealing relative angular changes of 10^{-7}–10^{-8} radians and relative interplanar spacing differences of 1 part in 10^7 or 10^8. The two crystals necessary can either be separate crystals (Brádler and Lang, 1968) or be different parts of one crystal (Bonse and Hart, 1965; Hart, 1968; Schwuttke and Brack, 1973).

6.1.5 Image Recording and Statistical Noise

By far the most usual method of recording the image is the photographic film or plate. Any photographic emulsion is sensitive to X-rays but it is important to choose the most suitable ones for topographic work. Table 6.1 shows the most commonly-used materials in order of decreasing

Table 6.1. *Photographic emulsions used in X-ray topography*
(In order of decreasing speed and increasing resolution)

Emulsion	Resolution (μm)	Thickness (μm)
Crystallographic X-ray film	~100	~20
Dental X-ray film	~10	~20
Ilford nuclear emulsion G5	1	10–100 as required
Kodak high-resolution plate	0.5	5
Ilford nuclear emulsion L4	0.25	10–100 as required

speed. The resolution required will be the main factor affecting choice of emulsion. Even if low resolution is required the slower plates will give superior images due to lower statistical noise, but the factor of several hundred times difference in speed between the fastest and slowest emulsions will mean that the fastest plate or film giving enough resolution will be used. The thinner emulsions (10 μm nuclear plate or spectroscopic plate) will be used for Berg–Barrett or other methods in which the plate is not placed perpendicular to the diffracted beam, to avoid loss of resolution. Dental or even crystallographic film is adequate for work such as the study of sub-grains or mask patterns on integrated circuits while the finest nuclear emulsion is used for top-quality transmission topography. This emulsion is available in several thicknesses; thin emulsions give higher resolution but are less efficient in stopping photons. To absorb half the incident radiation, one needs about 12 μm of emulsion for CuK$_\alpha$,

50 μm for MoK_α and 100 μm for AgK_α. The thicker emulsions are difficult to process uniformly and as a compromise either 25 μm or 50 μm emulsions are used. Processing must be carried out with great care, and at a low temperature to obtain uniform development throughout the depth of the emulsion (details of processing and of other factors affecting photographic quality will be found in the review by Lang, 1970). Although the grain size of the best emulsions is only about 0.25 μm, a single photon will cause a number of ionisation events over a range of about 1 μm with CuK_α and 2 to 3 μm with AgK_α. This is probably the most important factor limiting resolution, especially with harder radiations.

There have been several attempts to avoid the problems of low speed and long processing of photographic emulsions by the use of an image converter to display the topograph directly onto a television screen. Possible image converters include X-ray sensitive Vidicon or Plumbicon tubes, TV cameras utilising a dense array of silicon diodes (these methods depend on the photoconductive or photoelectric effects induced in certain materials by the X-ray beam) and intensification of the image formed on a fine-grain fluorescent screen. To date these have been quite successful in producing low-resolution images, comparable to those produced on dental film, at considerable cost; an achievement that does have significant applications such as on-line examination of integrated circuits for quality control. There is little doubt that a high-resolution image converter could be developed given enough effort. However, the statistical noise limitation discussed below means that to collect enough photons for sufficient resolution and image contrast in high-resolution work, the image converter would have to be exposed just about as long as a photographic plate. The only advantage would be that one could see the picture as it was being formed and reject badly set experiments or poor crystals at an early stage. Only synchrotron radiation provides a sufficiently intense source for one to consider performing dynamical experiments on a topographic camera monitored by a television system.

We have already encountered a statistical noise problem in scanning electron microscopy (section 2.4.2) where we found that a certain minimum number of electrons was required to provide a given resolution at a given contrast level. There is no need to repeat the calculation: we simply note that the situation is better for X-ray topography than for scanning electron microscopy, since on the scale we are working and with the emulsions in normal use, we may assume that the crystal scattering and photographic detection processes are not noisy. This results in an improvement of the final figures by about an order of magnitude and in figure 6.11 we have replotted figure 2.18 taking this into account and also altering the scales to show the number of photons required per square micron versus feature size or resolution, for different contrast levels. Lang (1970) reaches similar conclusions from a slightly different approach, and points out that the maximum number of photons that can be recorded per square micron on the best plates is about 8 for MoK_α radiation and 25 for CuK_α. These levels are shown on the graph, and the intersections of these lines with the curves show the minimum sizes of

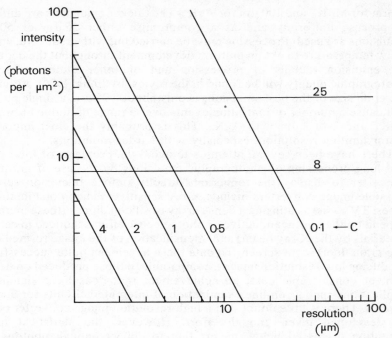

Figure 6.11. The number of photons per square micron required in order to resolve features of a given size at contrast levels from 0.1 to 4. The lines at 8 and 25 photons per square micron are the maximum intensities that can be recorded on nuclear emulsions with MoK$_\alpha$ and CuK$_\alpha$ radiation respectively

features detectable at various contrast levels. At low levels of contrast this is clearly an important limitation: collection of more photons merely makes the emulsion too black to see through (so a positive advantage of a high-resolution image converter would be that higher resolution at low contrast could be obtained by electronically making the system much slower than the slowest photographic plate!). This effect is seen in the higher magnification topographs in this chapter such as figure 6.8b; the apparent 'granularity' is not at all due to the photographic grain size around 0.25 μm but to the fundamental limitation of the statistical fluctuations in the generation of photons in the X-ray set.

6.2   CONTRAST OF IMAGES IN X-RAY TOPOGRAPHY

As with electron microscopy an understanding of the diffraction and absorption processes occurring inside the crystal is essential, not only to make full use of the methods but also to avoid positively misleading conclusions, since image contrast can change dramatically with experimental conditions such as the wavelength of the radiation. Again, we shall first study the diffraction effects in the perfect crystal and then look at the influence of imperfections. The majority of this section will be given to transmission effects since these are both more complicated and more informative than the reflection effects.

6.2.1 X-ray Diffraction in Perfect Crystals

Most students will be familiar with the geometric or kinematical theory of X-ray diffraction from experience with Laue and Debye–Scherrer methods. In X-ray topography however, we are generally dealing with thick, fairly perfect crystals. This is the very situation in which the simple theory breaks down, and it is necessary to use the dynamical theory of diffraction, which takes into account the rediffraction of once-diffracted beams. The mathematical treatment in the X-ray case is unfortunately considerably harder than in the electron case discussed in chapter 5, because electromagnetic waves are vector, not scalar waves and Maxwell's electrodynamic equations and field notation should be used (references to complete reviews of this very elegant theory will be found at the end of this chapter). However, in order to understand the main features of topographic images it is sufficient to use the conclusions of the electron diffraction theory of the last chapter, suitably modified for the X-ray case.

Most of the features of the theory of electron diffraction hold for X-rays, such as the excitation of two Bloch waves by a single incident wave, the interference between Bloch waves, the dispersion surface representation, the very low absorption experienced by the Bloch wave with nodes at the atomic planes (the Boormann effect) and the relation between the extinction distance and the dispersion surface diameter. There are, however, some important differences.

(1) The two-beam approximation is very much better for X-rays since the rocking curve is far narrower. Even the 'systematic' reflections are absent, and in practice it is quite difficult to set up multiple-beam X-ray diffraction with monochromatic radiation.

(2) Since X-rays are vector waves the state of polarisation is important. Different dispersion surfaces are found for polarisation vectors parallel and perpendicular to the incidence plane and for an unpolarised incident beam four Bloch waves rather than two will be excited in the crystal. For many purposes this effect may be ignored by assuming that the incident wave is polarised in one of these directions.

(3) The refractive index of solids for electrons is slightly greater than unity but for X-rays it is slightly less than unity. This only means that X-ray wave vectors are slightly shorter inside the crystal than outside and that it is the Bloch wave from the branch of the dispersion surface further away from the origin that experiences the lower absorption rather than that from the nearer branch as in the electron case. However, we label the low-absorption branch as branch 1 in both the electron and the X-ray cases since Bloch waves from these branches have the same symmetry (nodes at atomic planes).

(4)   The commonly used X-ray wavelengths are much longer and Bragg angles are in the range 5–45° rather than around 1° as in the electron case: this does have important consequences as will be seen below.

(5)   The assumption that the incident wave is plane is much worse for X-rays than for electrons.

An incident wave can only be approximated to a plane wave for the purposes of diffraction theory if its divergence is much less than the width of the rocking curve of the crystal, about 10^{-5} radians in the X-ray case. If a point source X-ray tube is used with, say, a 100 μm collimating slit, the source–slit distance must be 100 m if the beam divergence is to be 10^{-6} radians. The intensity at this distance would be minute (quite apart from the inconvenience) and in practice the divergence would be limited rather by the need to separate $K_{\alpha 1}$ and $K_{\alpha 2}$, at say 5×10^{-4} radians. A very low divergence can realistically be attained only by the use of a multiple-crystal monochromator.

The effect of spherical incident waves is easily understood by means of the dispersion surface representation (Kato, 1960). In chapter 5 we saw how a plane wave will excite one pair of tie-points on the dispersion surface, one on each branch, at positions determined by the angle of incidence and by the orientation of the crystal surface. A plane wave at a different angle will excite tie points at different positions. It follows that a spherical wave will excite tie points all along both branches of the dispersion surface. In a sense, the whole dispersion surface over the angular range of the incident beam is illuminated (figure 6.12a). Because of the scale of the dispersion surface the directions of the diffracted beams will be very little different from those expected from geometric theory. However the Bloch waves are very much affected. The direction of propagation of a Bloch wave is also the direction of flow of the energy carried by that wave (the Poynting vector) and, as shown by Kato (1958 and 1960), this direction is given by the normal to the dispersion surface at the excited tie-point. As seen in figure 6.12b, therefore, a whole range of Bloch waves is excited in the crystal and the energy flow is contained in a fan bounded by the incident and diffracted beam directions. This is called the 'X-ray fan' or 'Borrmann fan' and has an angular width of $2\theta_B$ (where θ_B is the Bragg angle). This is a real physical effect, and interference phenomena can occur (for example at defects) at any point inside the fan. Only at the exit surface do the Bloch waves in the fan split up into the 'incident' and 'diffracted' beam directions, and it is better to call the beam emerging in the incident direction the 'forward-diffracted' beam.

This effect can, of course, occur in the electron microscope if the condenser lens is set to give a very divergent beam. However, there are two important differences. The extinction distance is much longer for X-rays: this is inversely related to the diameter D_1D_2 of the dispersion surface, hence the dispersion surface is much more sharply curved in the X-ray case and it is much easier to excite the whole fan. Secondly, as mentioned above, the Bragg angles are very different. Even if a whole fan were excited in the electron case, it would only have a width of a couple

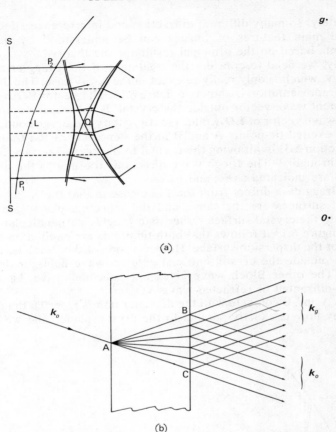

Figure 6.12. The effect of spherical incident waves on the excitation of Bloch waves: (a) reciprocal space; the divergent incident beam has incident wave vectors ranging from P_1O to P_2O; the whole dispersion surface over this range is illuminated, leading to the Borrmann fan ABC in (b) which shows the Bloch waves excited in the crystal and the beams generated outside the crystal

of degrees, which would not affect the accuracy of the column approximation for the calculation of image contrast even in the thickest specimens which can be used. In the X-ray case, however, the fan can easily be 60° in angular width and in a typical specimen could be hundreds of microns across at the exit surface. This clearly invalidates any column approximation for calculating the image contrast on a scale comparable with the resolution; it is the essence of the column approximation that the diffracted beams leaving a region of a certain diameter on the exit surface arise only from diffraction effects in a column of crystal of the same diameter.

The problem of calculating image contrast in such situations is so difficult that it has only been solved mathematically in a few simple cases.

Fortunately, so many different diffraction and interference effects occur that the main features of images can be understood by qualitative arguments based on the principal results of the theory.

Finally, we need to consider the treatment of the Bragg or reflection geometry, which is only really relevant in the X-ray case. The dispersion surface representation is shown in figure 6.13. The vector **PO** represents the incident wave vector outside the crystal. L is the Laue point and an outside wave vector of **LO** would exactly satisfy the Bragg condition. We find the excited tie-points A and B in the same way as the transmission case (section 5.3) by drawing the normal to the crystal surface SS which passes through P. The Bloch wave directions (normal to the dispersion surface) are indicated as b_A and b_B.

The Bragg case differs from the Laue case in that the X-ray entrance and exit surfaces are the same, and the reflecting planes are roughly parallel to the crystal surface rather than roughly perpendicular to it. As seen in figure 6.13 it follows that both tie points are excited on the same branch of the dispersion surface. However, one of the Bloch waves, b_B is directed outside the crystal and can give no wave fields or diffraction effects. The other Bloch wave gives one reflected wave **Ag** and one forward-diffracted or refracted wave **AO**.

When A and B coincide with the diameter points D_1 or D_2 the direction of energy flow is exactly parallel to the crystal planes and we have an

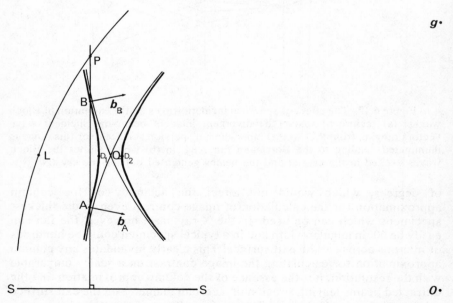

Figure 6.13. The dispersion surface representation in the Bragg case. **PO** is the incident wave vector, SS the specimen surface and A and B the tie points. The Bloch wave from B would be directed outside the crystal and is not, therefore, excited

analogous situation to the Borrmann effect. At orientations in between these points the normal to the dispersion surface passes in between the two branches of the dispersion surface and no tie points at all are selected. This range corresponds in a non-absorbing crystal to total reflection of the incident wave and since the Bragg condition is exactly satisfied when P coincides with L it is seen that this condition actually occurs at an angle slightly displaced from the 'true' Bragg angle. The displacement is dependent on the refractive index of the material, that is the distance LQ compared with QO in figure 6.13. To scale, if LQ is drawn as 10 mm then the distances QO and Qg become about 10 km, and the displacement from the Bragg position is only a few seconds of arc. In an absorbing crystal the reflection is not total, but still usually higher than in the Laue case; this is very useful for constructing efficient crystal monochromators. In any case the waves existing inside the crystal are very strongly attenuated (this is the phenomenon of primary extinction, mentioned earlier) and have negligible intensity after penetrating a few microns of crystal. In transmission methods, crystals several millimetres thick can be studied but reflection methods give information only about the surface layers.

6.2.2 Absorption, and Distribution of Intensity in Incident Beams

The contrast observed in X-ray topographs depends on the absorption as well as the diffraction properties of the specimen. There are several mechanisms of absorption, the strongest being the photoelectric absorption with emission of electrons. The total absorption can be expressed by the well-known equation

$$I = I_0 \exp(-\mu t) \tag{6.7}$$

in which I_0 is the incident and I the transmitted intensities, t the crystal thickness and μ the linear absorption coefficient. This equation is the basis of X-ray radiography since μ varies from element to element; relatively large defects or foreign objects in metals (or humans) can be detected by their different absorption properties. Some examples of values of the coefficient μ are shown in table 6.2 on page 232; it does not vary in a simple manner with wavelength because of the discontinuous changes at absorption edges (see for example Barrett and Massalski, 1966) but the dimensionless coefficient μt can always be used to characterise the absorption strength in a given material for a given radiation: figure 6.14 shows the graph of I/I_0 against μt from equation 6.7. In conditions appropriate to topography values of μt range between about 0.05 to 25.

There are three generally recognised types of image in transmission topography. Of course, these merge into each other but we can generalise by saying that direct images are predominant when $\mu t \ll 1$, intermediary images when $\mu t \sim 1$ and dynamical images when $\mu t \geqslant 5$. It is easier to understand the images if we first obtain a clear idea of the intensity profiles in the incident beam as it traverses a crystal. Figure 6.15 shows

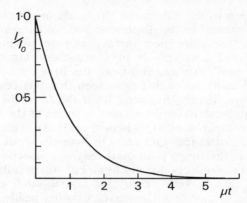

Figure 6.14. Graph of the fraction of the incident intensity that is transmitted, I/I_0, against the absorption thickness μt

these profiles as plots of I against θ, the angular spread of the beam. It is assumed that the crystal is set so that a reflecting plane is at the appropriate angle for maximum reflection from the centre of the profile in figure 6.15a. The curves reflect the fact that the angular spread in the incident beam is much larger than the width of the rocking curve of the crystal (section 6.2.1).

Several important points come out of these profiles.

(1) When μt is low the diffracted beam from the perfect crystal is weak, for either branch of the dispersion surface. However, the total intensity in the direct beam is high and imperfect regions can give strong diffraction since they may be able to reflect the whole angular range of the direct beam. Direct images result.

(2) When μt is ~ 1 the direct beam is still reasonably strong (about 35 per cent of the initial value) but the diffracted beams from both branches of the dispersion surface are also relatively strong. Interference can occur between all of these and the very complicated intermediary image results.

(3) When μt is large (figure 6.15d) the direct beam has negligible intensity. At the exit surface, the Bloch waves split up into a diffracted and a forward-diffracted beam, of approximately equal intensity. With this very high absorption, only the waves with nodes exactly on the atomic planes will get through so only branch 1 of the dispersion surface need be considered. The so-called dynamic or Borrmann images result and as explained in section 6.2.4 these are directly analogous to the thick-crystal case in electron microscopy. Direct and intermediary images are completely absent, but defects in the path of the Bloch waves disturb the perfect transmission along atomic planes and cast shadows on the image. Thus, imperfections give a weaker diffracted (or forward-diffracted) beam when μt is large whereas they give a stronger diffracted beam (and slightly attenuated direct beam) when μt is small.

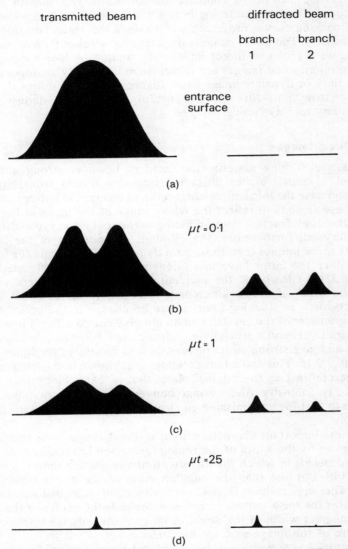

transmitted beam

diffracted beam

branch 1   branch 2

entrance surface

(a)

$\mu t = 0.1$

(b)

$\mu t = 1$

(c)

$\mu t = 25$

(d)

Figure 6.15. Schematic intensity profiles of incident and diffracted beams after traversing various absorption thicknesses of perfect crystal: (a) at entrance surface, (b) after $\mu t = 0.1$, (c) after $\mu t = 1.0$, (d) after $\mu t = 25$. The scales are the same in each figure; the (vertical) intensity scale is in arbitrary units, and the full width of the horizontal (angular scale) axes on the left-hand figures is about 2 minutes of arc

It is perhaps useful to compare the situation with that in electron microscopy. Electron scattering is much stronger than X-ray scattering, and the rocking curves are much wider (since the extinction distance is much smaller). The direct beam is therefore far weaker than in the X-ray case (figure 6.15b) and direct images are unique to X-ray topography. Electron microscope images are therefore most closely analogous to the intermediary or dynamic image cases (figure 6.15c or d) minus the direct image contribution; this gives a useful clue to the nature of the intermediary and dynamical images.

6.2.3 Direct Images

In weakly absorbing material the incident beam is strong and quite divergent. A much higher diffracted intensity results from imperfect regions because the local misorientations and changes in lattice parameter cause these regions to reflect the whole range of the incident beam, not merely the small fraction corresponding to the rocking curve width of the perfect crystal; furthermore, the off-angle components of the incident beam are more intense than those exactly on the Bragg angle for the bulk crystal since the latter have lost intensity to the main diffracted beam (figure 6.15b). Clearly, if the incident beam divergence were reduced below the width of the rocking curve there would be no direct image.

Defects such as stacking faults cause no change in the orientation or lattice parameter of the crystal, and do not give rise to a direct image. The elastic strains around a precipitate or dislocation, however, do cause such changes and give strong direct images, as was illustrated by figure 6.8 for which $\mu t = 0.15$. This and all other topographs shown in this chapter have the same contrast as the original plate; that is, extra blackening means extra X-ray intensity (the normal convention for X-ray topography, opposite to that for transmission electron microscopy as used in chapters 4 and 5).

The most important characteristic of a direct image is its width. This will be given by the width of the region (projected in the direction of the diffracted beam) in which the misorientation is greater than the rocking curve width but less than the angular width of the direct beam (figure 6.15b). The upper limit is not very important; for dislocations and precipitates the misorientations increase towards the centre of the defect and the angular width of the direct beam controls only the intensity near the centre of the image and not the width.

The actual image width can be calculated by an essentially geometric method (Chikawa 1965; Authier 1967, 1970), if the atomic displacements u caused by the defect are known as a function of position. If g is as usual the reciprocal lattice vector of the reflection, k the magnitude of the incident (or diffracted) wave vector and θ_B the Bragg angle then the effective misorientation is

$$\delta(\Delta\theta) = -\frac{1}{k \sin 2\theta_B} \frac{\partial(g \cdot u)}{\partial k_g} \tag{6.8}$$

the operator $\partial/\partial k_g$ signifying partial differentiation in the direction of the diffracted beam k_g. The displacements u are well known from elasticity theory in the case of a dislocation (the core region is here quite irrelevant) and it is straightforward, though a little tedious, to calculate the effective misorientation in the incidence plane and project this in the direction of the diffracted beam. An example is shown in figure 6.16 (Miltat and Bowen, 1975). In many cases the calculation predicts a double image as indicated here. This can often be observed but frequently the image is too narrow and the resolution insufficient.

The width of the image decreases as the rocking curve width increases. The rocking curve width is increased by increasing the wavelength and by decreasing the Bragg angle (using a lower order reflection, such as 111 instead of 224 in silicon, for example) and both these methods increase

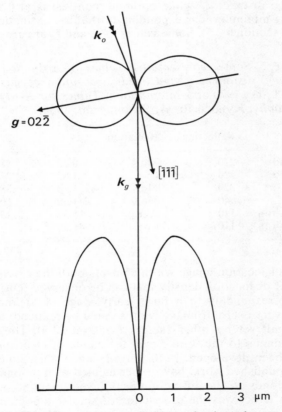

Figure 6.16. Contours of equal effective misorientation around a screw dislocation, showing also the diffracted beam direction k_g and, at the bottom of the figure, the expected image profile. Calculated for silicon, $b = \frac{1}{2}[\bar{1}01]$, $g = 02\bar{2}$, MoK$_\alpha$

the resolution and the upper limit of dislocation density which may be observed. On the other hand, the sensitivity to small strains increases as the rocking curve width decreases. Direct images are the narrowest and least sensitive to strains of the three types of image (typical values for dislocation image widths would be 1–10 μm) and are used when the dislocation types and configurations rather than their strain fields are being studied; this may not be possible when relatively thick or strongly-absorbing specimens have to be used because very short wavelength X-rays are not readily available. There is also a lower limit on the thickness of crystal that may be used, since for very thin crystals there is no difference between the intensities predicted on either the geometric or the dynamic theories and the distorted region around a dislocation does not add enough intensity to give contrast to the image. The lower limit for dislocation visibility is found to be about one-third of the extinction distance ξ_g while optimum contrast is at $0.88\xi_g$ (the first Pendellösung minimum, corresponding to the first extinction contour as explained in section 6.2.5). Some values of μ and ξ_g are given in table 6.2.

Table 6.2. Some representative values of extinction (Pendellösung) distances ξ_g and linear absorption coefficients μ (From Lang, 1970 and International Tables for X-ray Crystallography, Kynoch Press, Birmingham, 1964)

Crystal	Reflection	Radiation	μ (mm^{-1})	ξ_g (μm)
Diamond	220	MoK$_\alpha$	0.22	48
Diamond	220	CuK$_\alpha$	1.62	27
Silicon	220	MoK$_\alpha$	1.50	38
Silicon	440	MoK$_\alpha$	1.50	60
Silicon–iron	110	AgK$_\alpha$	15.5	13
Silicon–iron	110	CoK$_\alpha$	41.5	5.7

Since the dislocation image width controls both the resolution and the upper limit of dislocation density that can be observed (currently around 10^4 mm^{-2}, corresponding to a fairly early stage of yielding), it would certainly be very useful if image widths could be reduced; at least to the resolution limit set by other factors (section 6.1.3). Unfortunately, a method analogous to the weak-beam technique in electron microscopy has not yet been developed. Both Petroff and Authier, and Bowen and Miltat (in unpublished work) have experimented with topographs taken at up to 200 seconds off the diffraction peak (itself about 50 seconds wide), but the main effect was the production of similar topographs at greatly increased exposure times; precipitate images were slightly narrower. This is not, of course, an exact experimental analogue of the electron microscopy case and it is possible that really narrow images could be obtained if parallel crystal-monochromated radiation were used (to avoid Guinier–Tennevin type images from the continuous radiation) and two or three counters were employed so that the specimen crystal could be set at a known misorientation from a multiple reflection.

6.2.4 Dynamic Images

A lattice defect such as a dislocation or stacking fault will tend to disrupt the flow of energy along lattice planes in the Borrmann fan, and since the Bloch wave from branch 1 of the dispersion surface is the only one to survive in a strongly absorbing specimen a shadow will be cast by the defect in the fan. The situation is analogous to that discussed in section 5.5 for electron microscopy; the effect of the defect is to transfer energy from the type 1 to the type 2 Bloch wave ('interbranch scattering') but the latter is strongly absorbed and does not reach the exit surface. The illustration of magnetic domains (figure 6.24a) contains several dynamic images of dislocations as well, and it is seen that their contrast is the reverse of that in figure 6.8. Since it is the Bloch wave itself that is perturbed, and this does not split into the diffracted and forward-diffracted beams until the exit surface, the same contrast is obtained from images formed with either beam. As the absorption becomes weaker so the two images become more nearly complementary.

Close examination of the dislocation images in figure 6.24a shows that they are not uniform along their length, but are sharper near one end. This is a consequence of the large angular width of the Borrmann fan in the X-ray case. The shadow of that end of the dislocation that is near to the X-ray entrance surface will be more blurred than that near the exit surface; this effect is accentuated by scanning the specimen through the beam, and a stationary topograph gives a clearer image. Even the sharpest part of the image is normally two or three times wider than the direct image.

The description 'dynamic image' arose historically because it is necessary to use the dynamical theory of diffraction to understand the image; on the kinematical theory the total diffraction would be quite negligible at such high values of μt. The nomenclature is rather unfortunate, partly because interpretation of the intermediary image also requires the dynamical theory, but also because still another image-type occurs that is often labelled 'dynamic'. This is caused by lattice distortions that are not large enough to cause interbranch scattering (and therefore do not give dynamic images in the sense described above) such as the longer-range strain field of a dislocation, or elastic distortions in a perfect crystal. In such cases the Bloch waves can curve to follow the lattice planes. However, they may then end up travelling in different directions to Bloch waves in other parts of the crystal and interference fringes may result in the image. Some progress has been made in computing the resulting images (see for example Penning and Polder, 1961; Takagi, 1962; Authier, 1967; Authier and Balibar, 1970; Taupin, 1964). This type of dynamic image is best given the additional description 'wave-field curvature type' when it is necessary to distinguish it from the 'Borrmann' image discussed in the first part of this section.

6.2.5 Intermediary Images

The complication of intermediary images arises from the fact that Bloch waves from both branches of the dispersion surface will be present as well as the direct beam. However, the principle is similar to that in electron microscopy discussed in section 5.5. Any severe displacement of the lattice such as a dislocation or stacking fault will cause interbranch scattering and consequent interference of the wave fields. It is as if a new surface were introduced inside the crystal. The two Bloch waves decouple into two diffracted and two forward-diffracted rays; these immediately re-enter the crystal and in general excite eight new Bloch waves. Detailed consideration of the boundary conditions shows that two of these Bloch waves will be parallel but will arise from different branches of the dispersion surface; they therefore interfere and the 'beats' between them show as fringes in the case of a stacking fault and as oscillatory contrast if the defect is a dislocation, as seen in figures 6.17 and 6.18a. Intermediary

Figure 6.17. Section topograph of a natural diamond crystal 1.5 mm thick in central region, wedge-shaped at edges. $g = 111$, MoK$_\alpha$ radiation. μt is a little under 0.5. The main fringe pattern, covering most of the image and reminiscent of wood grain, consists of Pendellösung fringes. Superimposed upon this are several fan-shaped fringe patterns, which are intermediary images of dislocations in the crystal (under the conditions of a section topograph the direct image of a dislocation, if visible, would be a single black dot at the head of the fan-like intermediary image). (Courtesy of A. R. Lang)

images are often quite complicated, as in addition to the above fringes the direct or dynamical image (or even both) may be present, and further interference effects may be caused by beats between the Bloch waves created at a defect and those created at the entrance surface which have not intersected the defect. The full complexity of the intermediary image as seen in figures 6.17 and 6.18a only appears in a section topograph; scanning tends to blur the fringes since different sets of fringes are produced as the defect passes through different parts of the fan.

Interference always occurs when Bloch waves coincide, either by meeting at the exit surface or by following identical paths through the crystal. The resulting patterns are called 'Pendellösung' (pendulum) fringes and are analogous to extinction contours in electron microscopy. They are formed in perfect crystals as well as in those containing defects, especially if spherical incident waves are used as is usual in X-ray topography. In a section topograph they can be seen even with flat parallel

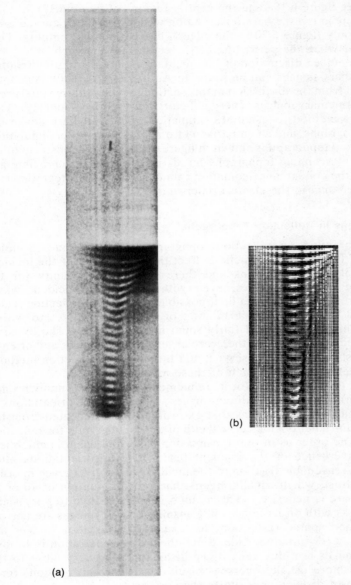

(b)

(a)

Figure 6.18. (a) Section topograph of an (001) slice of KDP (potassium dihydrogen phosphate), MoK$_\alpha$ radiation, μt approximately 0.5, $g = 200$. The vertical fringes are Pendellösung fringes and the complex pattern in the centre of the image is the intermediary image of a dislocation with Burgers vector parallel to [001], inclined at 20° to the crystal surface. (b) Computer simulation of Pendellösung fringe pattern and intermediary image of the dislocation topograph shown in (a). (Courtesy of E. Dunia, Y. Epelboin and J. F. Petroff)

crystals (see figure 6.18a and the central regions of figure 6.17), but in a scanning topograph they are observed on wedge-shaped or curved areas of crystal only (figure 6.20 at the edges), like extinction contours. They only appear when the perfection of the crystal is very high.

A more detailed discussion of the qualitative and quantitative features of intermediary images can be found in Authier (1967, 1970). As can be appreciated from the discussion in this section the quantitative interpretation of intermediary images is very difficult; however, Epelboin (1974) has developed a method for accurate simulation of intermediary images on section topographs, and a comparison of one of his computer simulations with an actual topograph is shown in figure 6.18. The only obstacle to the simulation of scanning topographs by his method is computer time and cost; since the column approximation cannot be used the integrations are far lengthier than in the electron microscopy case.

6.2.6   Images in Reflection Topographs

The effects found on reflection topographs are very much simpler. Because of the primary extinction effect the penetration of the incident beam is only a few microns and there is little opportunity for the interference of Bloch waves, even with a spherical incident wave. Furthermore, the crystal is set up if possible in the 'total' reflection region (section 6.2.1) in which the excitation of Bloch waves is zero with a non-absorbing specimen and fairly small in any material. The dynamic and intermediary images are therefore absent. Direct images and orientation contrast can both be observed, and the different reflection methods give different emphasis to each of these mechanisms.

The Berg–Barrett method (or its refinements using more monochromatic radiation) gives the best direct images. The spread of orientations in the vicinity of a defect means that extra intensity is reflected from the incident beam. The rocking curve width of a perfect crystal for reflection is of the same order as in the transmission case, and again is much less than the divergence of the incident beam. The misorientations must clearly be a reasonable fraction of the incident beam divergence in order to give contrast, which usually means that a misorientation of 10 or 20 seconds of arc is necessary. The method is not sensitive to very small strains but, as with direct images in transmission, the images are narrow and give good spatial resolution; for example, dislocation images are usually only a few microns wide. Since the X-ray penetration is so low, complete images are only seen if the dislocations lie very close to the surface and more usually one sees black dots where dislocations (or a dense group of dislocations) run to the surface.

Orientation contrast is a most useful feature of reflection topographs since it enables sub-grains to be studied without the restriction of the usual transmission methods which only image one sub-grain at a time. It also permits the study of orientation changes in quite heavily deformed materials, which are right outside the scope of transmission methods.

The study of orientation contrast can be made quantitative by observing the effects in different reflections (Wilkens, 1967). The possible misorientations of a reflecting plane can conveniently be described by rotations about three axes, A_1, A_2 and A_3 as shown in figure 6.19 (remembering, however that rotations do not add vectorially). A_1 is perpendicular to the reflecting planes (parallel to g) and rotation about this axis causes no contrast. A_3 is perpendicular to the incidence plane; a rotation γ about this axis will rotate the diffracted beam by γ, but if $\gamma > \Delta\theta$ (the width of the rocking curve) the diffracted intensity will be reduced to zero. A_2 lies along the intersection of the incidence plane and the reflecting plane, and a rotation β about this axis will cause a change of the angle of incidence approximately equal to $0.5\beta^2$. Thus if the rocking curve is a few minutes wide and β less than a few degrees, the reflecting plane will not deviate significantly from the Bragg position. However, the diffracted beam will be rotated (through an angle $\approx 2\beta \sin \theta$) about A_2 and an appreciable displacement will occur on the image.

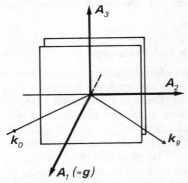

Figure 6.19. Definition of the axes of rotation for the discussion of orientation contrast in reflection topography. k_g and k_0 define the incidence plane; A_1 is normal to the reflecting planes, A_3 is normal to the incidence plane and A_2 is parallel to the intersection of these planes

Most of the variations of intensity will therefore be caused by rotations around A_3 but rotations about A_2 can be detected by displacements of any sizeable feature on the image. As with direct contrast the orientation effects will be controlled by the width of the rocking curve, $\Delta\theta$. As we have seen earlier, this is mainly determined in practice by the divergence of the incident beam; since a horizontal line source is usual for Berg–Barrett work, the divergence may be as much as one degree and it is necessary to reduce the divergence with slits if contrast from small misorientations is to be observed.

The above remarks apply to orientation contrast in any reflection method, but the sensitivity of the different methods to orientation contrast varies considerably, mainly through variations in the resolution of the system (affecting the minimum displacements that can be measured) and through differences in $\Delta\theta$. Both the Schultz and Berg–Barrett methods can be adjusted to give more or less sensitivity to orientation changes as explained in sections 6.1.1, 6.1.2, 6.1.3 and in this section, while the double crystal parallel setting (section 6.1.4) is sensitive to changes as small as a few tenths of a second of arc because in this case $\Delta\theta$ is very small. An interesting variation in this method is given if the second crystal is rotated very slightly about A_3 so that the final reflected intensity is about half the maximum possible; misorientations in the same sense as this rotation will then decrease the diffracted intensity whereas those in the opposite sense will increase it. The method can thus be made sensitive to the sense of strains in the specimen crystal.

6.3 SOME APPLICATIONS OF X-RAY TOPOGRAPHY

Many of the applications of this technique have been indicated during the discussion of image contrast but since the method is not as widely known as, say, transmission electron microscopy we shall conclude this chapter by summarising some of the investigations undertaken using X-ray topography.

6.3.1 Crystal Growth and Perfection

Industrial applications of high-quality crystals have increased enormously in recent years. The most striking example is certainly the field of semiconductor devices; these are produced from dislocation-free crystals of silicon or germanium since the presence of dislocations can ruin the performance of such devices. It was therefore essential to develop methods of growing perfect crystals that were large enough to ensure economic production. The problem was originally solved for crucible-growth methods by the etch-pit technique (Dash, 1959) in which a suitable etchant was used to produce etch pits at intersections of dislocations with the crystal surface, a method that can give the total dislocation density quite accurately but that gives little information about the dislocation distribution and none about the Burgers vectors or other detailed features of the analysis. The Lang method was developed in time to contribute to the problem of growing good crystals by the floating-zone technique (see Jenkinson and Lang, 1962) and was helpful in determining the mechanism of dislocation generation in this process.

In cases where the much faster etch-pit method gives all the information that is needed, topography can still be used in preliminary experiments to confirm that there is a one-to-one correspondence between etch pits and dislocations. Some etchants give etch pits where there are no dislocations, and others give pits for some dislocations, such as those 'decorated' with a Cottrell atmosphere of impurity atoms, but not others (for example, those freshly introduced by deformation). Silicon–iron is one metal for which a dislocation etchant has been successfully proved by topography; in this case by the Berg–Barrett method.

X-ray topography has been extensively used to study crystals grown both naturally and artificially, for example diamond (figure 6.17), quartz, potash alum, ice, sulphur, organic crystals such as the high explosive cyclotrimethylenetrinitramine, ceramics such as magnesium, aluminium and beryllium oxides and metals such as copper, aluminium and iron (for a review and references see Lang, 1970). In many cases the growth mechanism can be elucidated by studying the defects left in the crystal, for example by seeing whether dislocations are constantly generated during growth or whether the dislocations that formed in the nucleus and propagated through the crystal are the only ones present. Figure 6.20 is an illustration of the latter case, from the work of Duckett and Lang (1973)

Figure 6.20. Lang topograph of a whole crystal of hexamine (cubic structure) about 5 mm diameter and polyhedral. MoK$_\alpha$ radiation, $g = \bar{1}10$ and μt is much less than unity so only direct images are observed. (Courtesy of R. A. Duckett and A. R. Lang)

on hexamine. Only a few dislocations have been generated outside the centre and the hundred-odd dislocations that the crystal contains have clearly been created in the nucleus and propagated outwards with the growth of the crystal. In this and many similar studies, the ability of the method to reveal defects throughout the whole volume of a crystal is invaluable. The detailed study of the dislocations in this crystal also contributed to the confirmation of theories of crystal growth, since the deviations of the dislocations from straight lines radiating from the centre can be accounted for by the deflection of a dislocation from its original direction by the passage of a large growth step (formed from bunching of monomolecular steps) across the growth surface. The main object of the study of hexamine was to find a method of producing high-quality crystals of this material since it exhibits useful optical properties which improve with the crystal perfection (see for example Lee, 1969). The method used was slow, controlled evaporation of saturated solutions of hexamine in ethanol. Crystal growth from solution is particularly useful for organic compounds (which may decompose on melting) and after a number of unsuccessful attempts, monitored by X-ray topography, results typified by figure 6.20 were obtained. The crystal shown here contains a very low dislocation density, about 5 mm^{-2} and volumes of the order of 1 mm^{3} are completely free from dislocations; the high perfection of such regions is shown by the sharpness and regularity of the Pendellösung fringes which can be seen near the edges of the crystal where its profile is wedge-shaped.

6.3.2 Semiconductor Materials

The application of X-ray topography to the study of the growth of semiconductor crystals was mentioned in the last section. There have also been important developments in the study of semiconductor devices themselves or of the processes involved in their manufacture. A completed device has areas of pure silicon, areas covered by oxide or metal films and areas where the silicon is doped with an impurity element; Berg–Barrett reflection topography can be used as a quality-control check on the intermediate or final stages of the production process (see for example Rozgonyi et al., 1970) since these areas will have different diffracting power. The only problem that has prevented a wider use of X-ray topography in quality control has been the relative slowness of the photographic process. Although high-speed film and a rapid development process can be used if high resolution is not required, a total exposure and development time of some tens of minutes is still too slow for on-line quality-control at high production rates. The development of an efficient and reliable image converter (section 6.1.5) would be very useful in this context.

The steps involved in the production of integrated circuits are very amenable to study by X-ray topography. The first stage in the treatment of a silicon single-crystal slice is to steam-oxidise it at 1200°C. Windows are then etched into the oxide film and a suitable dopant such as phosphorus diffused through the windows by an anneal at, say, 1000°C in a flow of gas that contains the dopant. The dopant only penetrates a few micrometres into the silicon. Numerous problems can arise during these steps and in a badly controlled process the yield can be disastrously low. For example, dislocations can be introduced by thermal shock during the heating and cooling stages and the illustration of the Lang method in figure 6.8 is taken from a study of dislocation generation in a silicon slice (Miltat and Bowen, 1970; Miltat and Christian, 1973) treated as in the production of integrated circuits but with the doped layer etched away to show the dislocations more clearly. Such dislocations affect the performance of the circuit by locally enhancing the diffusion rate and distorting the impurity distribution which itself controls the electrical properties. Other problems can be caused by precipitation of second-phase particles (Schwuttke, 1962) and by dislocations generated in the diffusion process due to the lattice strain induced by impurity atoms of varying sizes (Schwuttke and Fairfield, 1966; Tanner, 1973).

6.3.3 Deformation Processes

The overwhelming majority of studies on the relation between crystal defects and mechanical properties has so far been made with the electron microscope, and X-ray methods cannot compete in the investigation of really small-scale phenomena such as the dislocation arrangements at high strains in work-hardening (or creep or fatigue), or the study of the interaction of dislocations with precipitate particles. On the other hand, the electron microscope is not so suitable for investigating either large-scale phenomena or the very early stages of yielding, and X-ray topography is beginning to make an impact in this field.

It is now possible to prepare many materials, including some metals, with dislocation densities low enough for transmission X-ray topography and a number of workers have constructed small deformation stages for the Lang or other cameras, (for example Young and Sherrill, 1971). Young and his colleagues at Oak Ridge have carried out much work on the deformation of copper crystals up to 1 mm thick; with this thickness only dynamic images can be seen, nevertheless the resolution is good enough to identify most of the dislocations in the specimen up to the yield point. Experiments have also been done on aluminium (for example Nøst and Nes, 1969), silicon (George *et al.*, 1973), copper whiskers (Nittono, 1971) and work has begun on body-centred cubic metals such as molybdenum (Becker, 1969) and silicon–iron. In some cases it is not possible to resolve individual dislocations in a slip band but one can still obtain useful information such as the internal distribution of slip bands, their nucleation points and the importance of secondary slip, as shown in figure 6.21 for the case of silicon–iron.

The specimen preparation and alignment is often quite difficult for deformation studies by transmission methods since the low dislocation density requirement (often only achieved in high-purity materials) means that the samples are very susceptible to accidental damage. For example, the copper specimens used by Young were so soft that they would bend under their own weight if picked up by one end. However, some very useful work can be done by reflection methods in which only the surface of a bulk specimen is examined. By applying short stress pulses to such specimens and measuring the distance moved by individual dislocations or by the tips of slip bands by a refined Berg–Barrett method, Vreeland, in a remarkable series of experiments has measured dislocation velocities as a function of stress and temperature in several materials (see, for example, Jassby and Vreeland, 1970). Unlike transmission methods, reflection methods may also be used at high strains. Figure 6.22, from the work of Coulon (1973) shows a Berg–Barrett topograph of an iron single crystal deformed by 30 per cent shear at room temperature. The long-

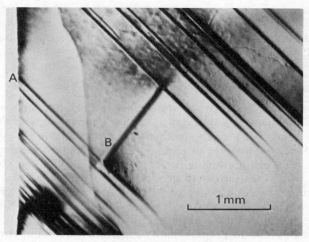

Figure 6.21. Lang topograph of slip bands in a deformed iron–2.5 per cent silicon alloy. Individual dislocations cannot be resolved in the bands. Note slip nucleation at specimen edge such as at A, and secondary slip at B. MoK$_\alpha$ radiation, $\mu t \approx 1$, $g = 200$. Stress axis vertical, stress ≈ 200 MPa. (Work of D. K. Bowen and J. Miltat)

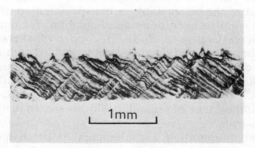

Figure 6.22. Berg–Barrett topograph of α-iron deformed 30 per cent at 296 K. CrK$_\alpha$ radiation, $g = \bar{1}00$. Macroscopic rotations over periods of about 170 μm are visible; the prominent lines running bottom left to top right are walls of dislocations on (111) planes, which form tilt boundaries of approximately 2° rotation. The wavy lines at right angles to the walls show the primary slip planes. Stress axis horizontal. (Courtesy of G. Coulon)

range rotations of the crystal and the walls of dislocations (revealed by analysis of the tilt across the observed boundaries) could scarcely have been revealed by any other method.

Even fracture can be studied by the topographic method. Figure 6.23 shows a Lang topograph of a brittle material (an α-sulphur crystal) containing cracks that have propagated by corrosion (Hampton *et al.*, 1974). The stress-relieving dislocations emitted from the crack tips are clearly seen and this technique could greatly extend the knowledge of

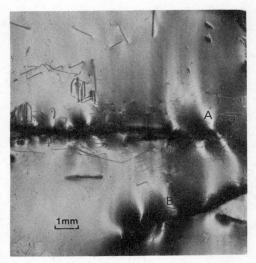

Figure 6.23. Lang topograph of α-sulphur, 1 mm slice, AgK_α radiation, $\mu t = 1$, $g = 222$. The strong black features are cracks; dislocations are seen to have nucleated at crack tips as at points A and B. (Courtesy of J. di-Persio and B. Escaig)

plastic blunting during brittle fracture, by providing precise data on the mechanism of crack propagation and on the work of fracture.

6.3.4 Magnetic and Electric Domains

Several methods are available for studying domain structures at the surface of a material, such as polarised light microscopy (the Kerr effect), particle decoration, scanning electron microscopy (section 2.5.6) but only X-ray topography and transmission electron microscopy are able to reveal domain structures in the interior of a crystal. Unfortunately, the magnetic domains occurring in thin foils, while interesting in themselves, are different from those occurring in bulk crystals and herein lies the usefulness of X-ray methods.

The contrast arises through the magnetostriction or electrostriction associated with each domain. In a single domain the magnetisation or polarisation will create a strain related to the direction of magnetisation or polarisation. For example, a specimen may basically have a cubic structure which on magnetisation changes to tetragonal, with the direction of the distortion parallel to the magnetisation direction. The distortion produced in ferromagnetic or ferrimagnetic domains is quite small, around 10^{-5} or 10^{-6}, but can be far larger in antiferromagnets or in ferroelectrics.

A crystal containing a number of such domains will therefore also contain extra strains due to the compatibility problem of fitting together a number of domains with the same basic crystal orientation but with slightly different structures. The orientation of each domain will be

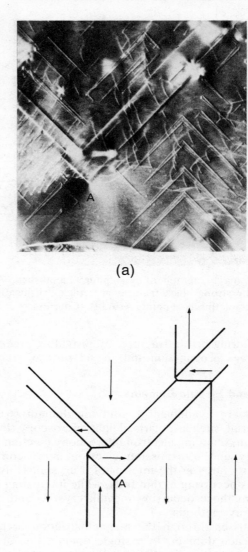

(a)

(b)

Figure 6.24. Magnetic domains in iron–2.5 per cent silicon. (a) Lang topograph, MoK$_\alpha$ radiation, $\mu t \approx 2$, $g = 020$; (b) diagram of the main domain walls and magnetisation directions. 180° domain walls are not visible in the topograph; several chevron-shaped surface domains can be seen. A domain wall junction at A shows a characteristic 'butterfly' image. The white streaks are dynamical images of dislocations. Field of view 1.4 mm × 1.5 mm. (Courtesy of J. Miltat and M. Kléman)

slightly altered to get the best fit and there may be more severe atomic displacements in the domain walls (where the structure is changing fairly sharply) or near the surface, due to surface relaxation. The contrast mechanisms can now be identified: (1) If the structure change is large, as in many ferroelectrics, the orientation of adjacent domains can be altered enough to give orientation contrast (this requires a rotation of the order of minutes of arc). (2) If the displacements in the domain wall or near the crystal surface are large, direct contrast will be seen, as in the case of a dislocation or precipitate, when μt is low. (3) If all displacements are small (as for example, in magnetic iron) the domains can still be observed since, as we have seen, the propagation of Bloch waves is extremely sensitive to small strains and strain gradients. Interbranch scattering occurs as the Bloch wave crosses the domain wall, giving contrast in the image as shown in figure 6.24a. This shows a junction of three domain walls at A; one is invisible because the magnetisation direction only differs in sense across the wall (a 180° domain wall) and the lattice parameter is therefore the same either side. The other two walls are clearly visible, as is the black–white 'butterfly', an additional effect caused by the combination of strains around the junction which is explicable in terms of disclination theory (Miltat and Kléman, 1973). Figure 6.24b shows the directions of magnetisation in the principal domains in this specimen.

Since domains move readily under the influence of an appropriate field it is possible to make slow dynamic studies as long as the domains are stable for the period of the exposure (Polcarová and Lang, 1962). These and all other dynamic studies will become far more fruitful when the powerful synchrotron radiation sources become more generally available.

6.4 GUIDE TO FURTHER READING

For further experimental details of the techniques the references given in the text or the detailed review by Lang (1970) should be consulted; Lang's paper also contains an extensive bibliography.

This chapter contains enough description of the theory for students to be able to understand most of the experimental work in this field and, in a qualitative way, its interpretation. Those proceeding to active research in X-ray topography must, however, come to grips with the dynamical theory of diffraction and its application to topography. A good procedure would be to read a brief review of the basic theory as given, for example, by Authier (1970); then to study one of the comprehensive reviews such as the excellent paper by Batterman and Cole (1964). An understanding of field notation and of the Maxwell equations is essential and can be obtained from the books by Feynman *et al.* (1963) or Shercliff (1976). For the application of the theory to X-ray topography, the reviews by Authier (1967 and 1970) are the best; these also give references to many of the original theoretical works. Useful papers are also to be found in various issues of *Advances in X-ray Analysis*.

REFERENCES

Authier, A., (1967), *Adv. X-Ray Analysis*, **10**, 9

Authier, A., (1970), in *Modern Diffraction and Imaging Techniques in Materials Science*, (eds S. Amelinckx, R. Gevers, G. Remaut and J. van Landuyt), North-Holland, Amsterdam, 481–520

Authier, A., and Balibar, F., (1970), *Acta crystallogr.*, **A26**, 647

Barrett, C. S. and Massalski, T. B., (1966), *Structure of Metals* (3rd edn), McGraw-Hill, New York

Barth, H., and Hosemann, R., (1958), *Z. Naturf.*, **13a**, 792

Batterman, B. W., and Cole, H., (1964), *Rev. mod. Phys.*, **36**, 681

Becker, C., (1969), *Phys. Stat. Sol.* **32**, 443 and **35**, 157

Bond, W. L., and Andrus, J., (1952), *Amer. Miner.*, **37**, 622

Bonse, U., and Hart, M., (1965), *Appl. Phys. Lett.*, **6**, 155

Bonse, U., and Kappler, E., (1958), *Z. Naturf.*, **13a**, 348

Brádler, J., and Lang, A. R., (1968), *Acta crystallogr.*, **A24**, 246

Chikawa, J. I., (1965), *J. appl. Phys.*, **36**, 3496

Coulon, G., (1973), *Docteur ès Specialité Thesis*, Lille

Dash, W. C., (1959), *J. appl. Phys.*, **30**, 459

Duckett, R. A., and Lang, A. R., (1973), *J. Cryst. Growth*, **18**, 135

Epelboin, Y., (1974), *J. appl. Crystallogr.*, **7**, 372

Feynman, R. P., Leighton, R. B., and Sands, M., (1963), *The Feynman Lectures on Physics*, Addison-Wesley, Reading, Mass.

Fiermans, L., (1964), *Phys. Stat. Sol.*, **6**, 169

Fujiwara, T., Dohi, S., and Sunada, J., (1964), *Jap. J. appl. Phys.*, **3**, 129

George, A., Escaravage, C., Schröter, W., and Champier, G., (1973), *Cryst. Lattice Defects*, **4**, 29

Hampton, E. M., Hooper, R. M., Shah, B. S., Sherwood, J., di-Persio, J., and Escaig, B., (1974), *Phil. Mag.*, **29**, 743

Hart, M., (1968), *Sci. Prog., Lond.*, **56**, 429

Jassby, K. M., and Vreeland, T., (1970), *Phil. Mag.*, **21**, 1147

Jenkinson, A. E., and Lang, A. R., (1962), *Direct Observations of Imperfections in Crystals*, (ed. J. B. Newkirk and J. H. Wernick), Interscience, New York, 471–95

Kato, N., (1958), *Acta crystallogr.*, **11**, 885

Kato, N., (1960), *Acta crystallogr.*, **13**, 349

Kohra, K., and Takano, Y., (1968), *Jap. J. appl. Phys.*, **7**, 982

Kohra, K., Hashizume, H., Yoshimura, J., (1970), *Jap. J. appl. Phys.*, **9**, 1029

Lang, A. R., (1959), *J. appl. Phys.*, **30**, 1748

Lang, A. R., (1970), in *Modern Diffraction and Imaging Techniques in Materials Science*, (eds S. Amelinckx, R. Gevers, G. Remaut and J. van Landuyt), North Holland, Amsterdam, 407–79

Lee, R. W., (1969), *Appl. Opt.*, **8**, 1385

Merlini, A. and Guinier, A., (1957), *Bull. Soc. fr. Minér. Cristallogr.*, **80**, 147

Miltat, J. and Bowen, D. K., (1970), *Phys. Stat. Sol.*, **a3**, 431; (1975), *J. appl. Crystallogr.*, to be published

Miltat, J. and Christian, J. W., (1973), *Phil. Mag.*, **27**, 35
Miltat, J. and Kléman, M., (1973), *Phil. Mag.*, **28**, 1015
Nakayama, K., Hashizume, H., Miyoshi, A., Kikuta, S., and Kohra, K., (1973), *Z. Naturf.*, **28a**, 632
Newkirk, J. B., (1958), *Phys. Rev.*, **110**, 1465
Newkirk, J. B., (1959), *Trans. Am. Inst. metall. Engrs*, **215**, 483
Nittono, O., (1971), *Jap. J. appl. Phys.*, **10**, 188
Nøst, B., and Nes, E., (1969), *Acta metall.*, **17**, 13
Penning, P., and Polder, D., (1961), *Philips Res. Rep.*, **16**, 419
Polcarová, M. and Lang, A. R., (1962), *Appl. Phys. Lett.*, **1**, 13
Rozgonyi, G. A., Haszko, S. E., and Statile, J. L., (1970), *Appl. Phys. Lett.*, **16**, 443
Shercliff, J. A., (1976), *Vector Fields*, Cambridge University Press
Schwuttke, G. H., (1962), *J. appl. Phys.*, **33**, 2760
Schwuttke, G. H. and Brack, K., (1973), *Z. Naturf.*, **28a**, 654
Schwuttke, G. H. and Fairfield, J. M., (1966), *J. appl. Phys.*, **37**, 4394
Tanner, B. K., (1973), *J. appl. Crystallogr.*, **6**, 31
Taupin, D., (1964), *Bull. Soc. fr. Minér. Cristallogr.*, **87**, 469
Takagi, S., (1962), *Acta crystallogr.*, **15**, 1311
Wilkens, M., (1967), *Can. J. Phys.*, **45**, 567
Young, F. W. and Sherrill, F. A., (1971), *J. appl. Phys.*, **42**, 230

7

Field–Ion Microscopy

One consequence of increasing the resolution of any technique is that the amount of material that can be studied at any one time decreases, since otherwise there is too much information present to be recorded or assimilated. Thus it would not be feasible to resolve simultaneously in the electron microscope all the atoms present in a thin foil. Where the electron microscope has been used to resolve atoms, either the planes of atoms in a crystal have been observed (essentially a diffraction effect), or else atoms deposited upon the surface of a film of some different material have been imaged. It is, in fact, almost essential to study only the surface of a solid at very high resolution, since the amount of detail to be recorded is thereby reduced to manageable proportions. One instrument that enables the atoms in the surface of a solid to be observed is the field–ion microscope, whose operation and advantages are described in this chapter.

7.1 DESIGN AND OPERATION OF THE MICROSCOPE

The instrument, shown schematically in figure 7.1, consists of an evacuated chamber, one side of which is a flat glass screen. The specimen, in the form of a sharp point with a radius of curvature of the order of 100 nm at the end, faces the screen. It is produced in this form by electropolishing a wire of the material in the manner indicated in figure 7.2 until it polishes away at the point where the wire enters the electrolyte (see Bowkett and Smith, 1970, §2.7). The end of the wire may then be sharp enough to produce an image, and it is mounted on the end of a cooled finger filled with liquid nitrogen or helium. The inside of the chamber is coated with an earthed conducting layer (usually tin oxide, which is transparent to light), while the specimen is connected to a variable high positive potential. The inside of the screen is coated with a suitable phosphor.

In use the chamber is first evacuated to a pressure below about 10^{-8} torr. An imaging gas, such as helium, is then admitted at a pressure of about 10^{-3} torr. When a sufficiently high potential is applied to the

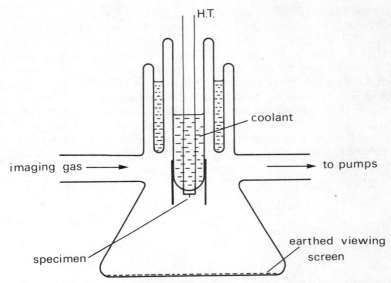

Figure 7.1. Schematic diagram of a field–ion microscope. The power supply raises the specimen potential to 10 to 40 kV, and can also heat the specimen by heating the resistive connection across the support leads

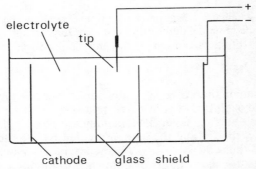

Figure 7.2. Schematic representation of bath for electropolishing the end of a wire to obtain a sharp point at the end

specimen, an image showing the positions of the atoms on the surface of the tip appears on the screen. The mechanism by which the image is produced is as follows. The gas atoms are polarised by the strong divergent electric field near the specimen and attracted to the tip, rather in the same way that iron filings are magnetised by a magnet and then attracted to the poles. When a gas atom reaches the tip it loses an electron in the very strong field there and becomes a positively charged ion. The electric field accelerates this ion radially away from the tip, causing the phosphor to emit light where the ion strikes it. Ions from different parts of the tip strike the phosphor at different points, so forming an image of the tip surface. The reason why atoms are visible in this image can be seen with the aid of figure 7.3. The surface of the specimen is not smooth but has steps in it where the planes of atoms intersect the surface. The local field depends upon the local radius of curvature, and so varies from point to point, being relatively high at points such as A, where atoms protrude and the local radius is small. Ionisation occurs preferentially at such points, so that ions stream away from them and the edge of a plane of atoms appears as a brighter region on the screen. Moreover, since this edge is not smooth, but is composed of a row of atoms, the edge is imaged as a row of bright points, each point corresponding to an atom. The image

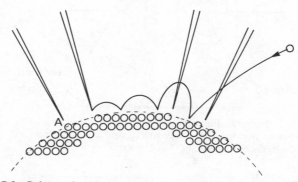

Figure 7.3. Schematic diagram showing the steps in the surface of the specimen due to the atomic nature of the material. Ions stream away from these steps to give rings of spots centred on the projection of a low-index crystallographic pole

of a good specimen therefore typically looks like the image shown in figure 7.4, where each bright spot corresponds to an atom that protrudes above the average surface level. The magnification is roughly equal to the distance between the tip and the screen divided by the radius of the tip, and is usually of the order of $10^6\times$. A plane intersects a sphere in a circle, and so the image consists of sets of concentric, roughly circular rings of bright spots, each set being centred on the projection of a low index pole, as indicated in figure 7.5.

If the radius of curvature of the surface is constant, and the ions forming the image stream radially away from the centre of curvature of the tip, the poles in the image correspond directly to part of a gnomonic projection of the specimen, as shown in figure 7.5. In practice, the tip is not hemispherical and the ionic trajectories are not radial, so the projection is distorted and often approximates to a stereographic projection. However, with experience the poles in the image may nevertheless be indexed and the orientation of the crystal determined. This is particularly useful since it enables orientation relationships across boundaries to be studied.

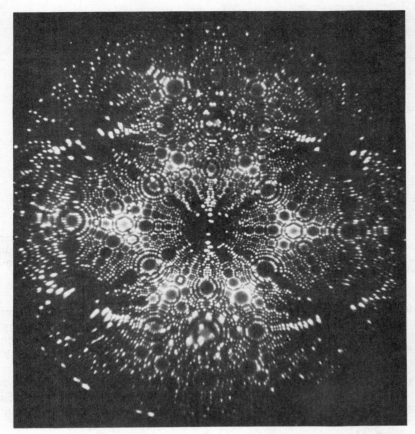

Figure 7.4. Field–ion micrograph of a tungsten tip, imaged with helium at 20 K. (Courtesy of T. J. Godfrey, D. A. Smith and G. W. Smith)

If the radius of curvature of the surface varies appreciably from one part to another, then the field will also vary and may be large enough to produce an image in some areas but not in others. Raising the voltage until all areas are imaged would seem to be the obvious solution to this problem, but there are two reasons why this does not work in practice. In the first place it is found, for reasons explained in the next section, that if the field is raised too far above the value that gives a good image, the contrast and resolution start to deteriorate. Raising the voltage may therefore cause the areas previously not visible to appear, but at the expense of the image quality over the rest of the specimen. Secondly, the forces that accelerate the ions to form the image also act upon the atoms that constitute the surface under observation. If the field becomes large enough then surface atoms in exposed sites, such as the image-forming rings of atoms, will be ionised and pulled out of the surface. This process,

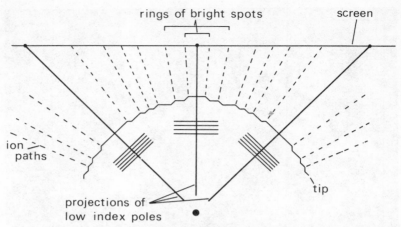

Figure 7.5. Diagram showing the equivalence of an ideal field–ion micrograph and a gnomonic projection of the tip: it is assumed that the ions travel radially away from the centre of the hemispherical end of the specimen (shown greatly enlarged)

known as field evaporation, tends, of course, to remove atoms preferentially from 'high spots', and is usually used as the final stage in producing a smoothly rounded tip suitable for microscopy. Since the field at which the best image is produced depends upon the material under observation, this problem is especially acute when attempting to image two-phase materials, since one of the phases may field-evaporate before the other gives a good image. In the case of pure materials however, field evaporation is generally useful in that one may thus remove the atoms layer by layer in a controlled manner (by pulsing the voltage applied to the specimen), and so reveal the three-dimensional nature of the microstructure under investigation.

The design of the microscope usually includes a cooled metal shield fixed onto the cold finger. This has the effect of cooling the gas atoms in the vicinity of the specimen, and so helping to keep the temperature of the specimen tip as low as possible. It also raises the density of the gas around the sample and thereby increases the supply of gas atoms to the surface. Many microscopes are made almost entirely of glass, these being straightforward to construct and bake out to give a good vacuum. On the other hand, in order to change the specimen or renew the screen it is necessary to break the glass at some point, and re-seal it again afterwards. Other microscopes commercially available are made of stainless steel, with the cooled specimen holder, viewing screen and pumping arm bolted on with standard high-vacuum connections. These are considerably more expensive, and being more complex can be subject to vacuum problems. They do have the advantage of versatility, however, since by providing the chamber with a number of extra ports it is possible to add on a variety of equipment necessary for *in situ* experiments.

7.2 PRINCIPLES OF IMAGE FORMATION

As explained in the previous section, the atoms of the imaging gas are attracted to the tip of the specimen, and therefore arrive at the surface with considerable kinetic energy. This surplus energy is given up, as a result of collisions, to the atoms in the surface of the specimen, which is usually cooled to as low a temperature as possible with either liquid nitrogen or liquid helium, since this improves its mechanical stability. Cooling the tip also increases the resolution of the microscope, because the random sideways motion of the ions as they leave the tip increases the spread of positions at which the ions from a given point on the surface arrive at the fluorescent screen. The colder the specimen the more slowly do the imaging gas atoms move over the surface of the tip when they have come to equilibrium, and the smaller their range of transverse speeds upon ionisation. The potential resolution of the instrument is given approximately by (Nishikawa and Müller, 1964).

$$\delta = 6 \times 10^{-3} \, Tr/F \text{ nm} \tag{7.1}$$

T is the tip temperature in K, r is the tip radius in nm and F is the field strength in V/nm. In general the resolution is not as good as this, but the temperature dependence is about right.

After an atom has arrived at the tip and slowed down it cannot escape because of the field, and moves about over the surface until it is ionised. It might be thought that the larger the field and the closer the atom to the surface the easier ionisation would become, but this is not entirely correct. Figure 7.6 shows the energy of the electron that is removed as a

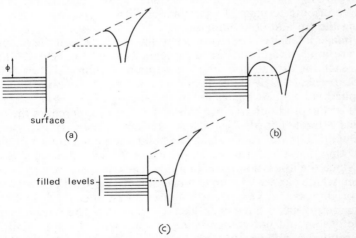

Figure 7.6. Schematic representation of the potential energy of the electron to be removed from the incident gas atom as a function of the electron position: (a) when the atom is some distance from the surface; (b) when the atom is close to the optimum distance; and (c) when the atom is too close to the surface, and there is no empty level into which the electron can tunnel

function of its position when the atom is at different distances from the surface. To escape from the atom the electron has to tunnel through the potential barrier formed by the atomic potential and the applied electric field. For an atom some distance from the surface the barrier has the form shown, and as the surface is approached the barrier becomes less, partly because the applied field increases, but also because the electron is attracted by the polarisation induced in the surface. Thus the ionisation probability increases as the atom approaches the surface until a critical distance is reached, after which the ionisation probability becomes extremely small. This is because the energy of the electron in the atom is, as indicated in the diagram, below the level of the Fermi surface, and so there is no vacant electron state in the solid into which the electron can tunnel. Consequently, there is an optimum distance at which ionisation occurs most readily, when the electron energy level is just above the Fermi level. Increasing the applied voltage reduces this optimum distance slightly, but it also increases the probability of ionisation for atoms at some distance from the surface. Therefore, raising the voltage above the value which gives a good image increases the proportion of ions that are produced some distance above the surface. These ions tend to have greater transverse velocities, and in addition, the field some distance above the surface is less dependent upon the details of the surface topography. As a result, ions produced at these levels carry less information about the atomic structure of the surface and since the proportion of such ions increases with the strength of the field, raising the voltage above the optimum causes the image quality to deteriorate. Impurity atoms are generally more easily ionised than the atoms of the imaging gas, and so normally are ionised an appreciable distance above the surface. Thus they do not contribute to image formation, and also they are prevented from reaching the tip and contaminating it.

The image intensity is determined by the rate at which the imaging gas atoms are ionised, and this, in turn, depends upon the rate at which the atoms reach the tip and upon the ionisation probability. At low fields, the ionisation probability is the controlling factor, and this varies rapidly with the applied voltage. At higher fields, the supply of gas atoms determines the rate of ion production; since raising the voltage increases the number of atoms attracted to the specimen, the image intensity increases with applied voltage in this range also. The helium ion current as a function of voltage thus has the form shown in figure 7.7 (Southon and Brandon, 1963). Since the ionisation probability depends sensitively upon the work function, which can vary by up to a fraction of an eV with orientation, the image brightness can vary from one area to another.

A number of gases have been used for imaging. The requirements are that the gas should not contaminate or react with the specimen, that it should require a sufficiently large field to produce a good image so that contaminants do not reach the surface of the specimen, and that it should give an adequate light output from the screen phosphor or image-intensification system. The commonly used gases are helium, neon and hydrogen. The imaging fields for a number of possible gases have been given by Müller (1965): typical values are shown in table 7.1.

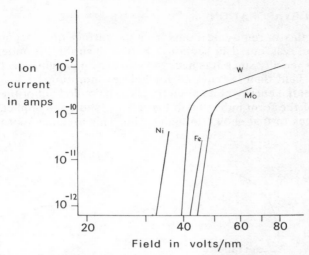

Figure 7.7. Helium ion currents from various metals as a function of the applied field at 78 K. Southon and Brandon, courtesy of *Phil. Mag.*

Table 7.1. Typical imaging
field values. (After Müller)

Gas	Field (V/nm)
He	45
Ne	37
H_2	23
A	23
Kr	19

The magnitude of the imaging field can affect the image in two ways. First, as the imaging field decreases so also does the resolution (equation 7.1). Second, with a given microscope a lower voltage has to be applied to the specimen when the imaging field of the gas is less, so that the total energy of the ions reaching the screen is reduced and the image intensity is lower. In addition the atoms with lower imaging fields have larger diameters than helium or neon, and in order to limit collisions between the gas atoms and the ions on their way to the screen, these gases have to be used at lower pressures. This also reduces the image brightness, since it reduces the supply of imaging atoms. It is to overcome these problems of image brightness that the image intensifiers described in section 7.4 have been devised. Helium, which does give a bright image, has the disadvantage of having a very high imaging field, restricting its use to the strongest materials, whereas the field for hydrogen is low enough to permit contaminants to reach the specimen surface, so that the initial vacuum in the system has to be very good.

7.3 FIELD EVAPORATION

The possibility of removing atoms from the surface of a specimen by field evaporation was noted in section 7.1. It is desirable to understand this process, especially since it is necessary to use an imaging gas that ionises at a lower field than is required for field evaporation.

The potential energy of an atom that is to be pulled out of the surface as an ion is of the form indicated in figure 7.8a, and upon application of the field changes to that shown in figure 7.8b. This assumes that the atom is

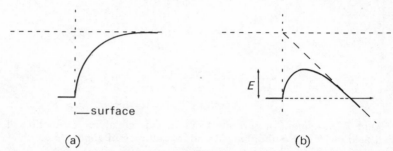

(a)   (b)

Figure 7.8. The potential energy of an atom in the surface of a solid as a function of its position (a) in the absence of any field; (b) with a large field applied, which ionises the atom as it leaves the surface

ionised as it leaves the surface. The potential energy of the ion depends upon the type of site it occupies in the surface, and the energy of the atom at A will be much higher than that at B in figure 7.9. When the field is applied, the atom at A will consequently have a smaller potential barrier to overcome than that at B, and will be removed preferentially. Its removal exposes the atom next to it at B, which is then more easily removed than before, and in this manner a whole layer of atoms can be stripped off.

The height of the barrier to be overcome by the ion under typical field

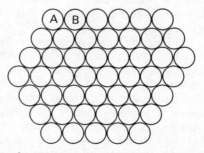

Figure 7.9. Schematic representation of atoms in the surface of a specimen: the atom at A has a higher energy than B (it is less strongly bonded) and so is more likely to be evaporated. Evaporation of the atom at A exposes the neighbouring atom at B, which is now more easily removed than before

evaporation conditions is about 0.1 eV, and its width is about 0.1 nm (Brandon, 1963, 1965). The ion can escape either by tunnelling through the barrier or, at higher temperatures, by thermal activation over it. The details of the process are evidently complicated, since it is observed experimentally that some of the field-evaporated ions are multiply ionised, but by making some simplifying assumptions the rate of field evaporation can be calculated (Bowkett and Smith, 1970, pp. 6–13; Müller and Tsong, 1969, chapter 3). Such treatments predict that the evaporation rate will depend strongly upon the applied field, so that changing the field by a few per cent may alter the evaporation rate by several orders of magnitude. Consequently, it is reasonable to define an evaporation field below which the evaporation rate is not measurable; above this value the rate increases rapidly. The observed value of the evaporation field is typically 40 to 50 V/nm for refractory metals, while many of the transition metals have fields in the range 30 to 40 V/nm. These values agree reasonably well with those calculated on the basis of the simple theory outlined above, which also predicts values of 16 V/nm for aluminium and 22 V/nm for tin. Since these values are low compared with the optimum fields quoted for various gases in the previous section imaging these materials in the field–ion microscope is not easy (see Boyes and Southon, 1972).

7.4 PRACTICAL PROBLEMS

As indicated in section 7.2, the main problem in applying field–ion microscopy to a particular material is that of getting an image of adequate intensity and resolution while keeping the tip free from contaminants. Were it not for the latter problem it would be possible to use hydrogen more frequently as the imaging gas, since it gives images of good brightness: however, its low field for best imaging does allow a number of contaminants to reach the surface of the specimen. Even when using helium as the imaging gas it is found that some contamination of the tip occurs, due to atoms diffusing along the wire from the surface of the cold finger. Thus it pays to use the imaging gas that has the highest imaging field that can be used without field evaporation occurring. An additional problem that arises is that, because of the fields employed, and hence the large stresses on the specimen, if the shank of the specimen contains a crack or dislocations it may break, leaving a tip whose shape is not suitable for microscopy. This prevents the direct examination of some types of specimen, such as cold-worked materials. The intensity at the screen is largely determined by the material under investigation, since this controls the choice of imaging gas. This, in turn, determines the potential to be applied to the specimen, since the best imaging field is quite closely defined. The pressure of the gas cannot be raised above a certain value, since otherwise the mean free path of the ions becomes shorter than the distance to the screen, in which case the ions suffer collisions with the other gas atoms and the resolution is impaired. The most frequently used imaging gases are helium and neon, but it has been found that mixtures of gases give better images under certain conditions. Thus 5 per cent of H_2 in He (Müller, 1967) gives greater brightness at reduced applied fields, while a helium–neon mixture can give better contrast and resolution.

However, the improvement that can be achieved is limited, and, even with extremely good photographic recording systems, the exposure times can range from a fraction of a minute up to half an hour or more. Therefore much effort has gone into devising ways of intensifying the image to improve the visibility for the operator and reduce the recording time.

An early solution to this problem was to modify the microscope design so that a larger overall potential was applied to the sample, thus increasing the energy of the ions reaching the screen, but also increasing the rate at which the screen phosphor deteriorated. Another approach was to place an ion-sensitive emulsion within the microscope itself (McNeil, 1968). Photoelectronic optical image intensifiers have also been used with success to amplify directly the light signal from the fluorescent screen (Müller, 1962).

The most useful approach at the present time seems to be that of replacing the viewing screen with a channel plate, which consists of a large number of very fine hollow fibres fused together to form a plate with the holes ('channels') in the fibres passing from one face of the plate to the other. A potential is applied between the faces of the plates, and ions striking one face produce electrons that are accelerated down the channels, producing more electrons as each electron collides with the walls of the channel, so that each channel acts like a photomultiplier tube. The electrons that emerge from the channels on the other face of the plate are focused magnetically onto a fluorescent screen, which can be viewed and photographed in the normal manner (Turner et al., 1972; Boyes and Southon, 1972). A gain in intensity of up to about 10^4 times is possible with this system, which also has the advantage of being much less sensitive to the imaging gas being used, so that gases such as argon can be used without difficulty. All intensification techniques involve a loss in resolution, but are in general capable of resolutions of about 20–40 line pairs/mm, which is adequate, especially at higher magnifications. Figure 7.10 shows a micrograph taken using a channel-plate intensifier.

7.5 APPLICATIONS OF FIELD–ION MICROSCOPY

Since the field–ion microscope gives atomic resolution of the surface of a small specimen, it is especially suited to the study at the atomic level of fine-scale microstructure that is normally uniformly distributed through-out a bulk sample. The size of the sample makes it impractical to search for detail which occurs only occasionally in macroscopic samples. Interpretation of the image is relatively straightforward when the specimen is a perfect single crystal of a pure material. Departures from perfection of a single crystal give rise to displacements in the regular spot pattern of the field–ion micrograph, and these can usually be interpreted in terms of the defects in the crystal. When the material contains more than one type of

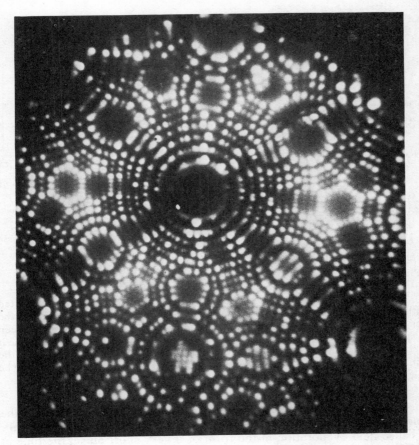

Figure 7.10. A field–ion micrograph of a tungsten specimen similar to that used for figure 7.4, taken with He gas at 78 K using a channel-plate intensifier: this reduced the exposure time by a factor of about 10^4. (Courtesy of T. J. Godfrey, D. A. Smith and G. W. Smith)

atom it may not be possible to distinguish between them on the basis of the contrast that they produce. A technique that enables atoms to be distinguished is to add a time-of-flight mass spectrometer to the microscope in the manner shown in figure 7.11, in which case the microscope is usually called a field–ion atom-probe. The image on the screen is now viewed from the same side as the field–ion specimen, since the space on the other side of the screen is occupied by a long (1–2 m) drift tube with a high-amplification electron multiplier tube as a detector at the far end. There is a small hole in the centre of the viewing screen, and the cold finger on which the specimen is held is flexibly mounted so that the image of any of the atoms in the tip can be brought into coincidence with the centre of this hole. The ions forming the image and coming from this atom

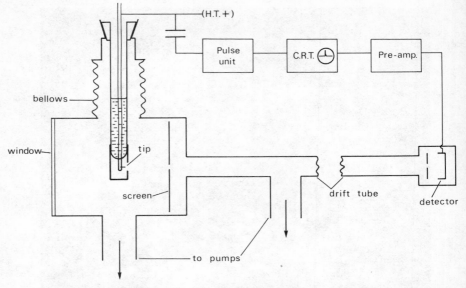

Figure 7.11. Schematic diagram (after Müller) of a field–ion microscope fitted with a velocity analyser (the field–ion atom-probe). The specimen is tilted until the image of the atom of interest coincides with the hole in the viewing screen. It is then evaporated as an ion by a short voltage pulse applied to the sample and the time it takes to reach the detector measured electronically. This gives the velocity of the ion, and knowing its kinetic energy from the applied voltage, enables its mass to be calculated

thus pass through the hole, and also the atom itself passes through the hole when it is field-evaporated. By applying the extra voltage necessary to cause field evaporation in the form of a very short pulse, the time at which the atom leaves the tip is known, and the time it takes to reach the detector can be measured. This, knowing the distance to the detector, gives the velocity of the ion and since its energy is known its mass can be deduced. By selecting in this manner the atom in a given area which experience tells is to be the next to be evaporated, and repeating the process a large number of times, all the atoms in a given region of the tip may be analysed. This technique is especially useful in, for example, precipitation studies.

7.6 POINT DEFECTS

The field–ion microscope is a particularly suitable tool for use in problems involving point defects, since currently no other technique is capable of imaging defects of this type. The instrument has to be used with care, however, since a number of problems arise that could lead to spurious results. A small burnt-out area on the screen phosphor could be misinterpreted as a vacancy, but the presence of this 'vacancy' in a large number of micrographs should cause its true nature to be recognised. In the absence of any image intensification the exposure times for recording the image may be long, and the film will suffer from reciprocity failure. This gives a threshold effect, so that atoms giving only a weak image are not recorded at all. Gradual field evaporation of the specimen will bring such atoms into positions where they should image strongly, and their presence or absence is then clear. Similar effects occur when a vacancy lies just below the surface, since the adjacent surface atoms are then displaced below the surface into the vacancy, so that they do not image. Again field evaporation should enable this situation to be recognised, since at some point the displaced atoms should be imaged, and in addition another vacancy image, due this time to the vacancy itself, is found in the next layer of atoms. Interstitials beneath the surface can similarly cause surface atoms to be imaged much more strongly than they otherwise might. An impurity atom may be misinterpreted as a vacancy, since it may be much less efficient at causing ionisation than the other atoms in the surface. In this case, field-evaporating atoms will bring the vacant image point to the edge of its plane, where the imaging intensity is relatively high, and the impurity atom may then produce a weak but discernible image.

The other, related problem which arises when using the field–ion microscope to study vacancies and other point defects is to find the defects in the image. A typical micrograph will contain of the order of ten true vacancies, and it is a problem finding them among the 10^4 or more normal image points, quite apart from detecting cases of spurious contrast. In cases where the density is high, as for example, after the specimen has suffered radiation damage, it is easier to find defects, but just as difficult to be sure that they have all been found. It would seem reasonable to suppose, however, that if defects are sought carefully the number detected, while not equal to the number actually present, will be an approximately constant fraction of the true number, so that the microscope can be used for comparative, even if not absolute, studies.

Since the careful study of a single layer of atoms on the surface of a specimen requires taking a number of micrographs, with part of an atomic layer being field-evaporated between each, it is desirable to have some means of distinguishing between those atoms that are present in consecutive micrographs and those that are not. This will reduce the amount of material requiring detailed study on each micrograph. Müller (1960) has devised an ingenious technique for doing this, in which images of consecutive micrographs are independently projected onto a screen so that the images are in register. One image is projected through a red and the other through a green filter, so that atoms that are present in both appear as yellow–white. Those present in only one of the micrographs appear as either red or green spots, so that the atoms that were evaporated and do not appear in the second micrograph can be seen and distinguished from those atoms that gave no contrast in the first micrograph but appeared in the second. This technique is also useful in studies of *in situ* radiation damage, where atoms are displaced between one micrograph and the next, and of surface diffusion, where the positions of the mobile surface atoms can readily be found.

The main application of the field–ion microscope in this area has been to the problems mentioned above: the vacancies present after quenching a specimen, the defects produced by radiation damage, and the movement of impurity atoms over the surface. By mounting the specimen on a wire through which a current can be passed, it can be heat-treated between micrographs and the annealing behaviour of the defects observed.

7.7 DISLOCATIONS

Since specimens made from deformed materials possessing a high dislocation density shatter in the high fields acting on the tip in the field–ion microscope, it is perhaps surprising that dislocations have been observed. Even if the specimen remained unbroken under these forces (which approach the theoretical strength of the material), it might be expected that the dislocations would be constrained to move out of the material to the surface and so be lost.

In view of these large forces, and also the small volume of material under investigation, the field–ion microscope is clearly not a suitable instrument for studying dislocation arrangements in deformed crystals or dislocation–dislocation interactions, except possibly where these are so large (that is, the dislocations so close) that the forces are greater than those imposed on the specimen by the field. Also, because of these forces, and the lack of sufficient resolution, it cannot give information about the positions of atoms in the core of a dislocation. However, using the atom probe described in section 7.5, the field–ion microscope is an obvious technique to use in studying the segregation of impurities and solute atoms in dilute alloys for dislocations and other defects.

A dislocation is visible at A in the field–ion micrograph shown in figure 7.12, where it can be seen that the effect of the dislocation is to cause the rings of spots encircling this point to be converted to a spiral. These rings of spots correspond to a set of lattice planes whose reciprocal lattice vector is g, and this conversion to a spiral results so long as the dislocation emerges at some point on the surface within the rings, and neither the line length l nor the Burgers vector b of the dislocation are normal to g (that is, $g \cdot b \neq 0$ and $g \cdot l \neq 0$). The reason for this can be understood with the aid of figure 7.13. In the absence of the dislocation the surface locally has the form shown in figure 7.13a, the roughly circular edges of the planes being imaged as rings of spots. The presence of a dislocation with Burgers vector b gives rise to the displacements shown in figure 7.13b. The component of the displacement normal to the planes which locally form the tip surface is $g \cdot b$ times an interplanar spacing which, since b and g are vectors of the real and reciprocal lattices respectively, is always an integral number of interplanar spacings. For a low-index set of planes $g \cdot b$ will usually be 1, 2 or perhaps 3, so that the vertical height of the step is 1, 2 or 3 interplanar spacings, and the planes match up across the join and are converted into a helical ramp. When the tip is smoothed by field

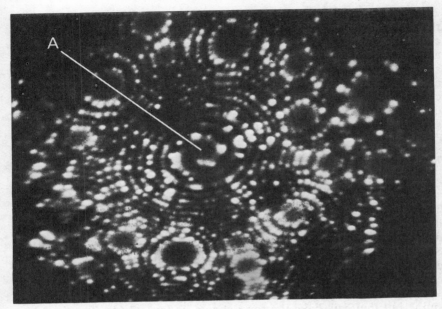

Figure 7.12. Field–ion micrograph of a sample containing a dislocation at A. (Courtesy of T. J. Godfrey, D. A. Smith and G. W. Smith)

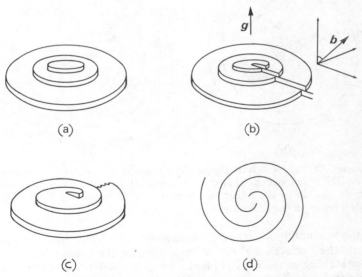

(a)

(b)

(c)

(d)

Figure 7.13. (a) Sketch of planes at the surface of a perfect tip; (b) effect upon these planes of introducing a dislocation with Burgers vector b such that $g \cdot b = 1$; (c) shape of the tip after smoothing, which converts the edges of the planes to a spiral; (d) form of the double spiral produced when $g \cdot b = 2$

evaporation it takes the form shown in figure 7.13c, which corresponds to the case $g \cdot b = 1$. When $g \cdot b = 2$ the result of the smoothing process is to cause the rings of spots to be converted to two interleaved sets of spirals as shown in figure 7.13d, each spiral giving an offset of one interplanar spacing. The direction of b determines whether the spiral is clockwise or anticlockwise. In situations where several dislocations are present, the configuration of a set of rings depends upon the total Burgers vector enclosed: thus a dislocation dipole will not displace the rings that surround both dislocations in the dipole.

7.8 PLANAR INTERFACES

7.8.1 Stacking Faults

The case of a stacking fault can be treated in a similar manner to that of a dislocation. A stacking fault that ends within a crystal does so at a partial dislocation, whose Burgers vector is not a vector of the crystal lattice. Suppose such a fault exists within a field–ion specimen, and the partial dislocation emerges through the surface at a point corresponding to the centre of a set of rings of spots in the image. Since $g \cdot b$ is not necessarily an integer, the displacements will not in general convert the rings to a spiral, but they will have the form shown in figure 7.14a. The line along

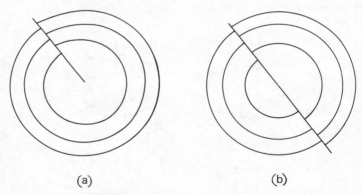

(a)   (b)

Figure 7.14. (a) Sketch of the spiral produced by a partial dislocation: the breaks in the rings occur where the associated stacking fault intersects the surface. (b) Sketch of the effect on the rings due to a stacking fault that is not terminated by a partial dislocation

which the mismatch occurs is the line along which the plane of the fault intersects the surface, and the mismatching of the rings corresponds to the mismatching of the crystal planes on either side of the fault. While in principle it is possible to determine the displacement R across a fault in the field–ion microscope, it is probably more readily accomplished in the electron microscope. When the fault crosses a set of rings, they take the form shown in figure 7.14b.

While the field–ion microscope may not be the most convenient instrument for studying stacking faults as such, it is, as in the case of dislocations, the only instrument for studying the details of segregation to defects at an atomic level.

7.8.2 Grain Boundaries

Low-angle grain boundaries may be thought of as an array of dislocations, and if these are sufficiently well separated they will be imaged individually in the micrograph. In order to determine the dislocation density, and hence the mismatch at the boundary, it is necessary to study the boundary at a range of depths in the specimen. This is because, as shown in figure 7.15, the number of dislocations observed at the surface depends both upon the density in the boundary and upon their inclination to the surface. A knowledge of the number of dislocations present, together with their inclinations and Burgers vectors, enables the misorientation to be calculated.

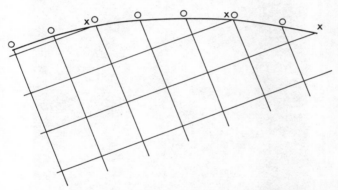

Figure 7.15. Schematic diagram of a small-angle boundary, showing that the local number of dislocations intersecting the surface depends upon their inclination to the surface. The two directions of dislocation shown have the same density, but one makes more intersections with the surface (marked with circles) than the other (marked with crosses)

High-angle boundaries present greater problems. It is relatively straightforward, using the fact that the image of a crystal looks approximately like part of its stereographic projection, to determine the orientation relationship between the grains on either side of a boundary. Ideally, it would be desirable to observe the positions of the atoms in and close to the boundary, and so obtain evidence to distinguish between various theories of grain boundary structure, but for several reasons this is not possible. In the first place, the displacements of the atoms from their perfect crystal sites are often small, and the instrumental resolution may not be sufficient to enable them to be measured. In the second place, the atoms in a boundary, lying in a region of 'bad' crystal, are less strongly bonded than usual. As a consequence, they are likely to be displaced under the influence of the strong electrostatic forces acting on them to positions that are not typical of their equilibrium configuration. Frequently, they field-evaporate more readily for the same reason, so that it is difficult to get a boundary region to give a good image. However, the field–ion microscope does seem to offer a method of observing segregation of impurity atoms to a boundary on an atomic scale. Auger and scanning auger spectroscopy and electron-probe microanalysis can also be used to study segregation, but give information about average levels of segregation over a relatively large area. An example of a grain boundary is shown in figure 7.16.

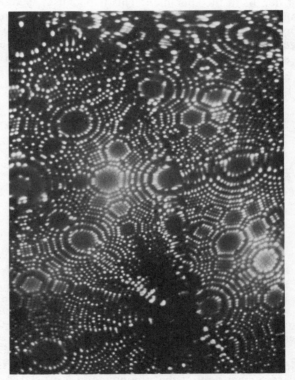

Figure 7.16. Field–ion micrograph of a specimen containing a grain boundary. (Courtesy of T. J. Godfrey, D. A. Smith and G. W. Smith)

7.9 TWO-PHASE SYSTEMS

In order to be able to apply the field–ion microscope to problems such as precipitation or ordering, it is necessary to be able to distinguish between one type of atom and another. This is straightforward if they give contrast that is clearly different, but can be done also, albeit less conveniently, using the field–ion atom-probe. Obviously many problems can arise in practice, so that not all systems of interest can be studied by this technique. An alloy may be difficult to form into a suitable tip, and may contain precipitates that cause it to shatter when the field is applied to it. The precipitates and matrix in an alloy may image at different fields, or one phase may field-evaporate before the field is sufficient to give an image of the other. However, the microscope can be used very fruitfully to study the initial stages of ordering and precipitation at the atomic level in suitable systems, and an example of ordering is shown in figure 7.17.

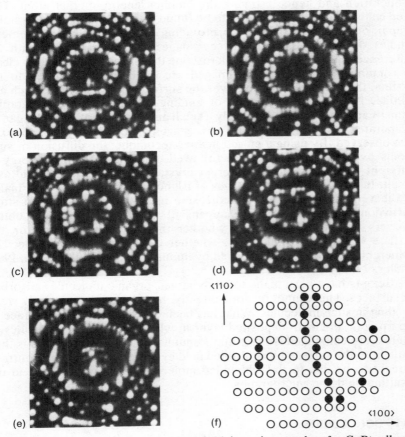

Figure 7.17. A sequence (a)–(e) of field–ion micrographs of a CoPt₃ alloy in which part of an atomic layer has been removed between each micrograph. Under the conditions used to produce the image the cobalt atoms are not visible, so that all the spots are due to platinum atoms. The local orientation corresponds to a {110} pole, and the diagram (f) shows (as dark spots) the extra spots observed in the micrographs and which correspond to platinum atoms on cobalt atom sites. From observations such as these the degree of order in the alloy can be determined. (H. Berg *et al.*, 1973; courtesy of *Acta metall.*)

7.10 SURFACE STUDIES

Since the field–ion microscope is essentially a tool for studying a surface, it is in the area of surface studies that its most rewarding applications are likely to lie.

As far as pure, clean materials are concerned, the field–ion microscope resolves the individual atoms in the surface, and is able to remove them one by one in a controlled manner by field evaporation. By measuring the voltage at which a particular atom evaporates, and knowing the shape of the tip locally, it is possible to calculate the local field that caused the evaporation and hence estimate the building energy of that atom. This then enables the binding energy to be found as a function of the crystallographic orientation of the surface, providing information that complements studies of the equilibrium surface shape of a material, but with the difference that in the field–ion microscope the surface is atomically clean. If contaminants that can be imaged are deliberately admitted to the system, then their distribution over the surface can be seen, (Ehrlich and Hudda, 1962) and their energy of binding to different crystallographic planes can be found. Alternatively a small amount of another metal can be evaporated onto the tip and the early stages of formation of an epitaxial film observed. By using a pulsed heating technique, the diffusion of such atoms over the surface can be followed, and the activation energy in different directions and on different crystallographic planes found as a result. In addition, the early stages of island growth can be seen. Finally, in the very important area of corrosion it should be possible to study corrosion or oxidation as it happens at an atomic level, and obtain answers to questions such as whether there are special sites on the surface where corrosion starts, whether certain planes are especially vulnerable, and what role is played by impurity atoms (Müller *et al.*, 1968; Müller, 1959).

Attempts have been made to study larger, organic molecules adsorbed on surfaces. This cannot be done directly, because of the relatively weak bonding in such molecules. What has been done is to coat the surface of the tip with an electrodeposited layer in which the organic molecules of interest are embedded (Müller and Rendulic, 1967). The molecule is then brought to the surface by controlled field evaporation, and extra contrast, thought to be due to the embedded molecules, is seen at points in the resulting field–ion micrographs.

7.11 CONCLUSION

The field–ion microscope, although it has been in operation for two decades, is still a young technique in the sense that it has never enjoyed the popularity of some of the others described in this book, due mainly to the practical problems that arise with many materials, and so has yet to realise its full potential. Undoubtedly within its limited range of operation it is a very powerful tool and can be expected to assist in solving many problems of commercial and scientific importance.

7.12 GUIDE TO FURTHER READING

The books by Müller and Tsong (1969) and Bowkett and Smith (1970), both give very detailed but readable accounts of this technique and its application. A rather more brief account by Müller is to be found in *Modern Diffraction and Imaging Techniques in Material Science.*

REFERENCES

Berg, H., Tsong, T. T., and Cohen, J. B., (1973), *Acta metall.*, 21, 1589
Bowkett, K. M. and Smith, D. A., (1970), *Field Ion Microscopy*, North-Holland, Amsterdam
Boyes, E. D. and Southon, M. J., (1972), *Vacuum* 22, 447
Brandon, D. G., (1963), *Br. J. appl. Phys.*, 14, 474
Brandon, D. G., (1965), *Surf. Sci.*, 3, 1
Ehrlich, G., and Hudda, F. G., (1962), *J. chem. Phys.*, 36, 3233
McNeil, J. F., (1968), *Metallog.*, 1, 91
Müller, E. W., (1959), in *Structure and Properties of Thin Films*, (eds C. A. Neugebaur, J. B. Newkirk and D. A. Vermilyea), Wiley, New York, 476
Müller, E. W., (1960), *Adv. Electronics Electron Phys.*, 13, 83
Müller, E. W., (1962), *Proc. Int. Conf. Crystal Lattice Defects*, Kyoto; (1963), *J. phys. Soc. Japan*, 18, Supp. II, 1
Müller, E. W., (1965), *Science*, 149, 591
Müller, E. W., (1967), *Surf. Sci.*, 8, 462
Müller, E. W., (1970), *Modern Diffraction and Imaging Techniques in Material Science*, North-Holland, Amsterdam
Müller, E. W., and Rendulic, K. D., (1967), *Science*, 156, 961
Müller, E. W., and Tsong, T. T., (1969), *Field Ion Microscopy*, Elsevier, New York
Müller, E. W., Panitz, J. A., and McLane, S. B., (1968), *Rev. Scient. Instrum.*, 39, 83
Nishikawa, O., and Müller, E. W., (1964), *J. appl. Phys.*, 35, 2806
Southon, M. J., and Brandon, D. G., (1963), *Phil. Mag.*, 8, 579
Turner, P. J., Cartwright, P., Boyes, E. D., and Southon, M. J., (1972), *Adv. Electronics Electron Phys.*, 33, 1077.

8

Unconventional and
New Techniques

This chapter describes fairly briefly a number of imaging techniques that are not widely used at the time of writing, either because they are rather specialised in their application or because they are relatively new and are not yet established. Since all of these techniques use electrons to form the image, in many cases the equipment is basically similar in design to microscopes described in earlier chapters. An instrument has been included in this chapter if it satisfies the criterion of producing contrast in a partly or wholly new way. For example, high-voltage microscopes are in a sense unconventional, in that their cost makes them relatively rare. Also, they tend to be used for studies in which their higher electron energy is either an advantage (for example, studying thicker materials to see the three-dimensional arrangement of the sub-structure) or a necessity (studying dynamical diffraction effects or producing *in situ* electron radiation damage). However, they are not considered in this chapter since there is no new principle or technique involved in producing contrast, although in practice the image contrast may appear to be different as a consequence of several strong reflections occurring simultaneously. On the other hand, the mirror electron microscope is included, since it produces contrast in ways not possible with other instruments.

Of the microscopes described here, the one that seems to have the greatest range of applications, and the one most likely to become widely used in the future, is the scanning-transmission electron microscope. This combines the versatility of a scanning microscope, that enables a number of contrast modes to be used and allows the image to be processed electronically, with the high resolution obtainable from a standard transmission microscope. Its high cost and the inconvenience of having to work in high vacuum will probably limit the spread of this instrument for the time being.

The field-emission microscope, the electron analogue of the field–ion microscope, is described first since it does not use any lenses to form its image.

8.1 FIELD-EMISSION MICROSCOPE

The field-emission microscope is the electron equivalent of the field–ion microscope, and is historically the earlier instrument. The basic design of the apparatus is very similar, the major differences being that the chamber is pumped to a high vacuum, since there is no need for any imaging gas, and the voltage applied to the specimen is negative instead of positive. The image is produced, not by positive ions originating at or just above the surface of the specimen, but by electrons that tunnel out of the surface and are then accelerated to the viewing screen by the field near the tip. There is no need to cool the specimen, since, as explained later, the surface temperature has a negligible effect upon the resolution. Indeed, one of the advantages of the technique is that the specimen may be studied at any temperature between that of liquid helium and room temperature or above.

As shown in figure 8.1 the specimen is usually mounted on a heating wire, so that the tip can be heated to clean it before use, as well as enabling observations to be made at high temperatures if desired. The temperature to which the sample can be raised is limited either by the evaporation of the surface-adsorbed atoms of interest or by the self-diffusion of the emitter material over the surface, which leads to blunting of the tip.

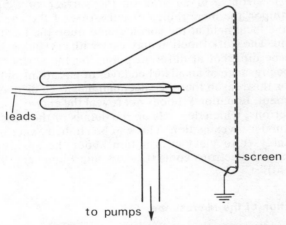

leads

screen

to pumps

Figure 8.1. Schematic diagram of a field-emission microscope. The chamber is pumped to a high vacuum and the specimen, in the form of a sharp tip, is mounted on a heater wire and faces the earthed fluorescent screen. The specimen can be annealed or cleaned by passing a current along the heater leads

8.1.1 Field Emission

The electrons forming the image have to tunnel through the potential barrier shown in figure 8.2. A general calculation of this tunnelling probability is extremely complicated, but at relatively low temperatures ($T \leqslant 600 \text{ K}$) the current density J due to tunnelling is given in A/m^2 by

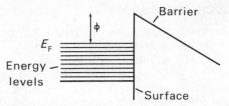

Figure 8.2. Representation of the energy barrier through which the electrons have to tunnel in order to escape from the metal and form the image. For those electrons at the Fermi surface the height of the barrier is approximately ϕ, the work function. Including other effects in addition to the applied field rounds off the top of the barrier and lowers it a little

(Good and Müller, 1956)

$$J = [1.54 \times 10^{-6} \, E^2/\varphi t^2(y)] \exp [-6.83 \times 10^{-5} \, \varphi^{3/2}/v(y)E] \qquad (8.1)$$

where E is the applied field at the surface in V/m, φ is the local value of the work function in eV and $t(y)$ and $v(y)$ are functions of $y(y = (Ee^3/\varphi)^{\frac{1}{2}})$. This equation has been verified for a number of materials, and shows that the current will vary very rapidly with the applied field E, and that it depends very sensitively upon the work function φ. Therefore the current density from a given area on the surface of a specimen will depend both upon the local radius of curvature of the specimen (which determines the local field at the surface) and upon the local value of the work function. The work function may vary with crystallographic orientation, and so be different at different points on the surface, and will be altered by the presence of an adsorbed layer of gas atoms or other surface layer. It is for this reason that the specimen has to be cleaned at the start of an experiment. Equation 8.1 does not reveal the energy spectrum of the escaping electrons, which depends on the details of the band structure of the material under investigation. This can be studied experimentally with a velocity analyser to yield information about the energy levels from which the emitted electrons come (Gadzuk and Plummer, 1973; Swanson and Bell, 1973).

8.1.2 Operation of the Microscope

The sharp needle-point tip used as the specimen is prepared by electrolysis in a similar manner to field–ion microscope specimens. In this case, the tip can have a larger radius since the field necessary for field emission is lower than that for field ionisation. It is usually smoothed to a suitable shape for microscopy, either by field evaporation or by annealing it and allowing surface diffusion to remove any irregularities. A crystal that has been allowed to come to thermal equilibrium is bounded by faces that are low-energy crystallographic planes, so that a specimen annealed in this manner has facets of low-index planes connected by rounded regions, as shown schematically in figure 8.3. Since the work function is substantially modified by adsorbed surface layers, it is essential to outgas or field-desorb the surface after the working vacuum has been attained.

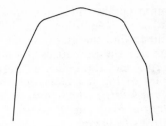

Figure 8.3. Representation of the shape of a field-emission tip after annealing

The electrons emitted from the tip under the influence of the applied field travel more or less radially from the surface to the screen, which therefore shows the variation in emission from place to place over the tip. The magnification is comparable with that of the field–ion microscope, but the resolution is appreciably worse, being about 2.5 nm at best. This is due to the fact that the electrons within the solid have kinetic energies of several eV if they are close to the Fermi surface, whereas atoms in equilibrium with the tip in the field–ion microscope may typically have an energy of 1/400 eV. Consequently electrons that tunnel through the surface have relatively high transverse speeds so that the spread of positions at which they strike the viewing screen is large compared with the field–ion microscope, and hence the resolution is lower. On the other hand, since the temperature of the solid has very little effect on the kinetic energies of the electrons within it, the microscope can be operated at room temperature without affecting the resolution. The emitted electron current is high compared with the ion current in the field–ion microscope, so the image is easily observed in a darkened room and there is no need to use sophisticated intensification equipment. Since the image is formed with lower fields at the specimen surface than those used in the field–ion microscope the mechanical strength of the specimen is less of a problem and does not usually limit the range of materials that can be studied. The usual difficulty is to obtain a clean surface, since it is not possible to flash non-refractory materials to a high temperature. Even so, materials such as copper, aluminium and silver have been imaged successfully, and even mercury, condensed on a cold tip in the form of whiskers, has been observed in the field-emission microscope.

8.1.3 Applications of Field-emission Microscopy

This technique can be used to study two types of problem: the intrinsic properties of clean surfaces, and the adsorption, diffusion and agglomeration of impurity atoms on an initially clean surface.

As far as clean surfaces are concerned, the most important quantity that the microscope can be used to measure is the work function φ, which depends upon crystallographic orientation. The current emitted by any part of the surface can be found, using a Faraday cage that collects the electrons that have passed through a hole in the viewing screen. Since the shape of the tip produced by annealing is not a hemisphere, the field at the surface varies from point to point and is not known, and so the measured current cannot be substituted in equation 8.1 to give a value of φ directly. In addition, it is difficult to find the emitted current density J precisely, since the irregular nature of the tip causes the magnification, and hence the relationship between J and the collected current i to vary over the surface. However, a plot of log i against the reciprocal of the applied voltage, gives a value for $\varphi^{3/2}/\beta$, where β is the field/voltage factor. If some other measurement is made, such as the half-width of the energy distribution of the electrons (this gives $\beta/\varphi^{\frac{1}{2}}$), β can be eliminated to give φ. Because of the tip shape, the variation in brightness of the image on the screen does not directly give even a qualitative estimate of the variation in work function across the surface.

As noted above, only materials that can readily be cleaned can be used directly in the field-emission microscope, but softer, more easily contaminated materials such as copper can be studied if they are grown by vapour deposition as epitaxial crystals on a tungsten tip within the microscope itself (Melmed, 1965, 1967). This not only enables their work function to be determined, but also allows the diffusion of these materials over the tungsten surface to be observed as a function of temperature and time, and hence activation energies for this process to be found.

This last type of experiment is really in the second category noted at the start of this section, that involving adsorbed atoms on the field-emitting surface. Clusters and layers of such atoms show up in the microscope image because they alter the effective work function of the surface and hence change the emitted current. The details of the mechanism by which this happens are not simple, and the reader is referred to the account given by Müller (1970). By using the contrast that arises in this manner, a range of adsorption and diffusion problems can be investigated. Even those gases with relatively high vapour pressures have been studied, since it has been found that immersing the whole microscope in liquid helium allows them to be adsorbed on the specimen surface (Gomer, 1961). Most of the work to date has used tungsten as the substrate since it is relatively easy to prepare good emission tips from this material.

An indication of the appearance of the micrographs and the information that can be drawn from them is given by the sequence shown in figure 8.4.

8.2 THERMIONIC AND PHOTOEMISSION MICROSCOPY

The major disadvantages of the field-emission microscope stem from the fact that very large fields are necessary to draw electrons out of the surface of the sample and the area under observation is very small. Producing the electrons by some other mechanism, such as by heating the material so that they are emitted thermionically, or by illuminating the surface so that photoemission occurs, enables these problems to be overcome, since now a very high field is not required, and therefore the surface can be flat and hence considerably greater in extent. The electrons emitted by either of these processes are accelerated as before by maintaining the specimen at a high negative potential. Since the surface is

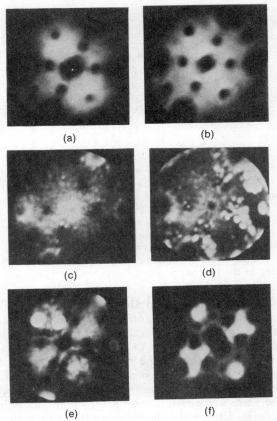

(a)   (b)

(c)   (d)

(e)   (f)

Figure 8.4. A sequence of field-emission micrographs: (a) a clean tungsten surface; (b) the same covered with a layer of nickel 3 atoms thick; (c) the same after the surface has been saturated with oxygen; (d) after heating at 680 K for 3 minutes the surface film has formed into crystallites, which appear as the bright circular areas; (e) after 1 minute at 1210 K the crystallites have started to disperse by evaporation and diffusion; (f) after 1 minute at 1700 K: the pattern is characteristic of oxygen on tungsten. These observations are correlated with changes in the work function which occur at each stage. (Courtesy of J. P. Jones)

now approximately flat, the geometry is such that the large magnification produced by simple projection in the field emission no longer arises, and it is necessary to focus the electrons onto the viewing screen by a set of magnetic lenses so that the microscope has the form shown schematically in figure 8.5.

Instead of a gun and condenser lens system to illuminate the specimen, the surface is effectively self-luminous as a result of its being at a potential of typically -30 kV, and being either heated (Kinsman and Aaronson, 1972) or illuminated with ultraviolet light of high intensity (Wegmann, 1972a, b). A screening ring may be used to improve the

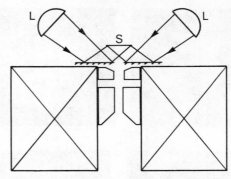

Figure 8.5. Schematic layout of the first stage of a photoemission microscope. The specimen S is maintained at a large negative potential, and is
illuminated by intense ultraviolet light from the lamps L reflected onto the
specimen by the mirrors located on the top of the objective lens polepiece. The
mirrors and the polepiece are earthed, and electrons emitted by the specimen are
accelerated through the hole in the mirror and focused by the lens to form an
image. Additional lenses magnify the image further, producing a final image on a
fluorescent screen at the bottom of the column. (After Wegmann)

geometry of the electrostatic field that accelerates the electrons, the
accelerating electrode at earth potential being the upper surface of the
objective lens polepiece. An objective lens aperture removes those
electrons travelling at a large angle to the optic axis, and so reduces the
effect of spherical aberration on the resolution. The rest of the image-
forming lenses enlarge the image produced by the objective lens, the
major difference compared with the transmission microscope being that
diffraction effects are not possible with a self-luminous object, and it is
therefore unnecessary to be able to focus on the back focal plane of the
objective lens. In this respect, the instrument is at a disadvantage
compared with the field-emission microscope, which, like the field–ion
microscope, shows the crystallographic orientation of the tip on the
screen directly.

Of necessity, the specimen is observed at a high temperature in the
thermionic emission instrument, but in the photoemission version the
sample may be heated or cooled if *in situ* experiments are to be carried
out. The other respect in which this type of microscope is inferior to the
field-emission microscope is that of resolution. Due to the aberrations
introduced by the electrostatic accelerating field the best resolution so far
attained is of the order of 20 nm. However, this resolution is better than
that of an optical microscope and while under most conditions it is no
better than that given by an SEM, the emission microscope is able to
produce contrast by mechanisms not possible with either of these
instruments.

As in the case of the field-emission microscope, contrast arises from
variations in the intensity of the emitted electron current. Assuming that
the temperature (or the light intensity, as appropriate) is uniform across

the specimen, variations in current will be due solely to variations in work function and those complications experienced with the field-emission microscope due to the variable field strength across the surface will be absent. The following factors can affect the effective work function and hence produce contrast.

(a) *Orientation.* This causes different grains in a polycrystalline material to appear with different intensities, and so be readily visible, even though the surface is featureless when observed by optical or scanning microscopy. This obviously has application to problems such as re-crystallisation or grain growth (Zaminer, 1968).

(b) *Composition.* For example, precipitates in an alloy will have a different work function from the matrix material, and so will be visible without etching (Graber *et al.*, 1968). Consequently phase morphologies can be observed and phase-transformation kinetics studied.

(c) *Oxide or corrosion layers.* Dielectric layers on the surface of a metal reduce the effective field at this point and so increase the distance the electrons have to tunnel in order to emerge into free space: varying thicknesses of oxide down to one atomic layer can thus be observed with great sensitivity.

(d) *Surface relief.* Some of the contrast due to surface relief may arise because, in a single-crystal grain, two regions that are tilted with respect to one another will often have different crystallographic orientations and hence different work functions. However, the major part of the contrast is caused by the fact that any irregularity in the surface will alter the local direction of the accelerating field experienced by the electrons: very close to the sample the field must be normal to the surface at each point. This effect gives contrast, even in the absence of orientation contrast (and hence from amorphous materials), and is so strong that it is not possible to get a useful image if the surface under study is too uneven.

These types of contrast, especially relief contrast, are similar to contrast mechanisms using the SEM but with the difference that the contrast is much greater with the emission instrument, especially when a surface film on a substrate is being observed. These differences are due to the greater influence of the work function on emission than on secondary electron production, and to the strong focusing action on the slowly moving emitted electrons by the irregularities in the field close to the surface.

Examples of the different types of contrast obtainable using electron emission are shown in figure 8.6.

8.3 ENERGY-ANALYSING AND ENERGY-SELECTING MICROSCOPES

These instruments are sufficiently alike in their basic principles to be considered together conveniently. They are derivatives of the standard transmission microscope, but modified to provide additional information about the specimen not given by the standard instrument. Their operation

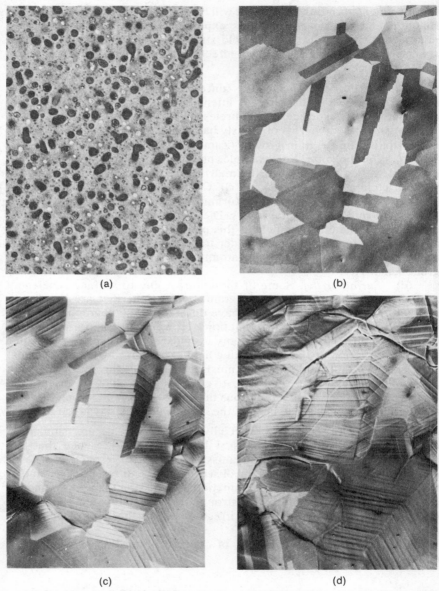

Figure 8.6. Examples of photoemission microscope contrast: (a) atomic-number contrast from precipitate particles in a glass (8000×: Courtesy of Balzers Ltd); (b) orientation contrast from sub-grains of different orientation in a copper specimen; (c) relief contrast at slip lines produced when the sample in (b) is deformed; (d) the same taken at 380°C after more deformation: relief contrast is again visible and the orientation contrast shows that much recrystallisation has taken place (500×: Courtesy of Balzers Ltd)

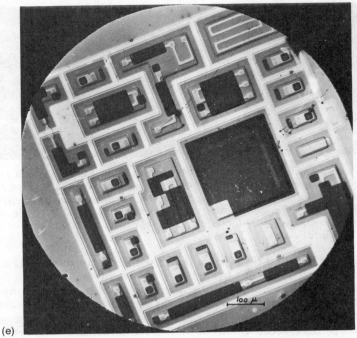

(e)

Figure 8.6 (contd.). (e) Atomic-number and relief contrast from an integrated circuit (100×: Courtesy of Balzers Ltd)

depends upon the fact that some of the electrons that pass through a specimen are inelastically scattered, and the energy-loss spectrum of these electrons is related to the electronic structure of the material of the specimen. Most substances give an energy-loss spectrum that contains more or less sharp peaks; these are known as the characteristic losses of the material.

8.3.1 Energy-analysing Microscope

By mounting an energy analyser beneath the viewing screen of a standard transmission microscope the loss spectrum of the electrons that pass through a slit in the screen can be found. A typical spectrum from an aluminium specimen (in which the strongest energy loss, due to plasmon excitation, is at 15 eV) is shown in figure 8.7b. The entrance slit of the analyser was set across the bend contour in the position indicated in figure 8.7a: the fringes can be seen not only in the line of intensity corresponding to those electrons that have not been inelastically scattered, but also in the lines that correspond to losses of multiples of 15 eV. This type of spectrum is obtained using either a Möllenstedt analyser (Cundy *et al.*, 1966) or a Wien analyser (Curtis and Silcox, 1971).

Since materials have their own more or less characteristic energy

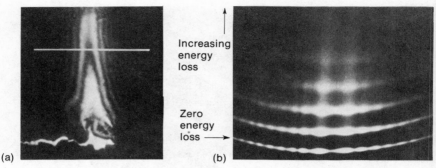

Figure 8.7. An electron energy-loss spectrum of magnesium obtained using a Möllenstedt analyser: (a) micrograph of the specimen showing the position of the entrance slit of the analyser; (b) the loss spectrum. It can be seen that the fringes visible along the line formed by the zero-loss electrons are also visible in the bright lines corresponding to those electrons which have excited one or more plasmons. (Courtesy of S. L. Cundy)

losses, it is evident that forming an image with electrons whose energies lie within a narrow range has two advantages. In the first place, magnetic lenses suffer from severe chromatic aberration and those electrons that have lost energy degrade the quality of the final image, since they cannot be brought to the same focus as the zero-loss electrons. If the analyser is set to accept only those electrons that have not lost any energy, the quality of the resulting image, especially when the specimen is thick, should be much improved. Secondly, if a specimen consists of grains of different phases, each with its own characteristic loss spectrum, and electrons corresponding to a characteristic loss of one of the phases are selected to form the image, then that phase will show up with lighter contrast against the general background.

Images of this sort can be obtained with the energy analysers mentioned above if a detector is mounted below the analyser to collect energy-selected electrons that have passed through a particular point in the slit. By scanning the image over this point in the slit and using the detector to modulate the intensity of a television monitor scanned in synchronism, an image can be built up. This is obviously not the most efficient method of producing an image, since most of the electrons forming the image on the fluorescent screen are not used at any one instant. However, because electron detectors are very much more sensitive than photographic emulsions, the exposure times are not inordinately long. Adding an energy analyser to the SEM enables similar contrast, but based on the differences in secondary electron spectra from different materials, to be obtained from that instrument.

8.3.2 Energy-selecting Microscope

A microscope that enables an energy-selected image to be obtained directly is the one devised by Castaing and Henry (1962), which uses a magnetic prism to obtain the energy dispersion. This prism is positioned in the microscope column between the objective and intermediate lenses, with the result that an image is formed in the normal manner on the viewing screen, but the electrons used to produce the image lie within a narrow band of energies. The layout of the components is illustrated schematically in figure 8.8. A standard objective lens forms a diffraction pattern of the object at the plane F_0, where the usual objective aperture is located. In the absence of the prism this lens would form an image of the object at I_0. The effect of the prism is to deflect the electrons through about 90° and to provide some energy dispersion. The electrons are then reflected by an electrostatic mirror, which consists of a suitably shaped electrode maintained at a slightly higher potential than the filament in the microscope gun. The electrons pass back through the prism where they suffer an additional deflection of about 90° so that they are again travelling along the optic axis of the instrument. This second passage through the prism produces additional energy dispersion. The plane F_1 is conjugate to F_0, so that the portion of the diffraction pattern admitted by the objective aperture appears at F_1, but is dispersed in energy. If the intermediate lens is focused on this plane, the energy spectrum can be seen on the final viewing screen (figure 8.9a). The electrons in this energy spectrum appear to come from a virtual image of the specimen in the plane I_1, so that if an

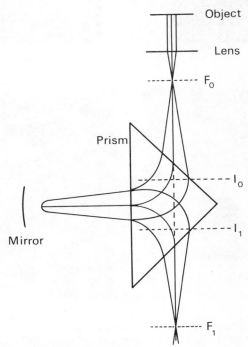

Figure 8.8. Schematic representation of the electron trajectories through the analysing prism in the Castaing–Henry energy-selecting microscope. (After Castaing)

aperture is inserted in the plane F_1, to select only those electrons within a small energy range, and the intermediate lens is focused on I_1, an image of the specimen formed using only these selected electrons will be produced on the viewing screen. (In the instrument designed by Henry, the dispersing prism was in fact located between the first and a second intermediate lens, but the description given above still applies except that the lens in figure 8.8 is the first intermediate lens, and its object is now the image produced by the objective lens. The plane F_0 is conjugate to the back focal plane of the objective lens, so that the diffraction pattern and the objective aperture are in focus in this plane.)

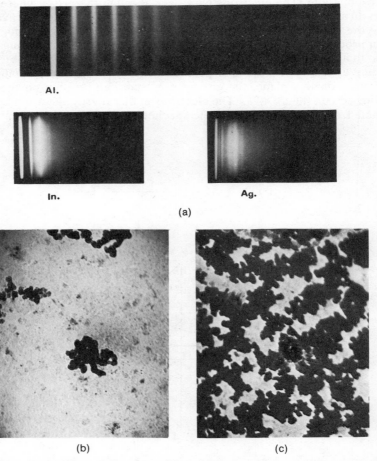

Figure 8.9. Micrographs taken with the Castaing–Henry microscope: (a) loss spectra of aluminium, indium and silver; (b) micrograph of an aluminium foil taken with zero-loss electrons; (c) the same area as (b) but taken with electrons which have lost 6.7 eV, corresponding to the oxide layer

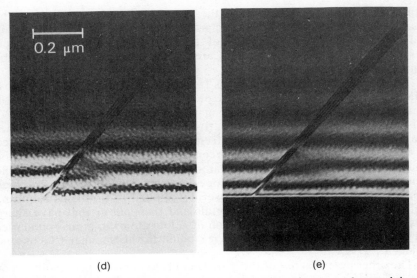

(d)   (e)

Figure 8.9 (contd.). (d) and (e) are micrographs of a crystal containing a stacking fault: the similarity in the contrast in the no-loss image (d) and the plasmon-loss image (e) is evident. (Spectra and micrographs courtesy of L. Henry)

In principle, the aberrations of the magnetic prism do not affect the resolution of the microscope to the first order. However, because of the limited dispersion of the prism, the finite size of the objective aperture increases the spread in energy of the electrons used to form the image. Consequently a compromise has to be sought between the conflicting requirements of high resolution in the image and good energy discrimination. In practice the problem is largely overcome since the effect of using an intermediate lens before the analyser is to demagnify the diffraction pattern and its superimposed image of the objective aperture at the plane F_0. This reduces the effect of the size of this aperture on the accuracy of energy selection, which in any case does not have to be better than 1 eV or so because of the spread in energies of the electrons leaving the gun. The resolution in the image formed using the zero-loss electrons is, except for the thinnest specimens, better than can be obtained from a standard transmission microscope. Since an objective aperture is used in the normal position, contrast can be obtained by the mechanisms considered in chapter 5, as well as that arising from using selected electrons.

This brief description does not do justice to an ingenious instrument, and other texts (such as Castaing, 1971) should be consulted for a more complete account.

8.3.3 Applications

To date, most of the applications of this type of instrument have been to problems concerned with the inelastic scattering itself. This has been stimulated both by advances in the theoretical understanding of the various types of inelastic scattering process, and also the desire to obtain information about the positions of the various loss peaks and their strengths. Work continues in this area since much still remains to be understood.

The general picture that emerges from this type of study is that most materials have characteristic loss peaks that are fairly broad and ill defined, the strongest peaks occurring in the energy-loss range of 5–25 eV. Thus it is difficult to distinguish clearly between one material and another on the basis of its energy-loss spectrum, in contrast, say, to using emitted X-rays, where the distinction between one element and another can be made very clearly. However, compared with the X-ray microanalyser, the energy-analysing type of microscope has the advantage of very high spatial resolution, and has been usefully applied to problems requiring this resolution, involving those metals and their alloys that have sharp loss peaks. For example, segregation of solute atoms to grain boundaries in dilute aluminium alloys has been studied in this manner (Silcox and Vincent, 1972).

The other potential use of the instrument is to improve the resolution of the microscope for thick specimens by removing the inelastically scattered electrons. Because many of the electrons emerging from a thick foil have lost some energy, only a relatively small fraction of those transmitted will be used to form the zero-loss image, which will in consequence be weaker, but sharper, than that obtained using a standard instrument. Since the inelastically scattered electrons give similar contrast to the zero-loss electrons, it would be desirable to be able to combine the contributions from these in some way to improve the overall intensity. The scanning transmission microscope, which is able to do this, without an appreciable loss in resolution, is described in section 8.5. Some examples of energy-selected micrographs are shown in figures 8.9b–d.

8.4 MIRROR ELECTRON MICROSCOPY

As noted in the previous section, an electrode will reflect a beam of monochromatic electrons falling on it at near normal incidence when the electrode potential is slightly more negative than that of the electron source. When the difference between these two potentials is small, the electrons almost reach the surface of the electrode before their velocity is reversed. If the surface is flat it acts effectively as a mirror, but if the field near the surface is disturbed in some way the electron trajectories are deflected. Since the electrons are moving very slowly in this region, they are very sensitive to such changes in the field, and are deflected appreciably by small perturbations. This effect can be used to study the surface of a material if it is used as the mirror electrode, and an image produced in which contrast arises by distinguishing between the deflected and the undeflected electrons. This can be done if an image is formed by a lens using the reflected electrons, and an aperture in the back focal plane of the lens removes those electrons which have been deflected. Alternatively, an image can be formed by what is in effect a point projection

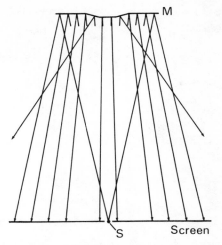

Figure 8.10. The point projection method of obtaining contrast. The source S in the centre of the screen illuminates the sample with a divergent beam of electrons. An equipotential surface just in front of the surface acts as a mirror M, the reflected electrons from a flat surface giving even illumination of the screen. Surface relief causes the electrons to be reflected in different directions and thus produces contrast on the screen

method, shown schematically in figure 8.10, although generally such instruments also have lenses through which the reflected electrons pass.

Contrast can arise from a variety of causes. If the surface is not flat, the reflecting field will be perturbed in the region of the irregularities, so that surface relief will be visible, essentially for the same reason that it can be seen with the emission microscopes considered in section 8.2. Again, the sensitivity of the technique is such that it is not possible to produce a good image unless the surface is very nearly flat. The reflected electrons are also sensitive to local changes in the electric potential of the surface. Thus insulating areas that become charged will be visible, and this effect can be used to study the current flow through microelectronic circuits, in a similar manner to the scanning electron microscope (section 2.6.3). The electrons are also sensitive to magnetic fields, so that the mirror micros-cope can be used to study the domain pattern at the surface of magnetic materials.

While the method of contrast production is in essence simple, the designs of those microscopes used to observe it are in general relatively complicated, especially when compared with, say, a scanning electron microscope. This is because the technique works best when the electrons are incident on the specimen surface at or near normal incidence, with the consequence that the reflected electrons return along the optic axis of the illumination system, and the final condenser lens is also the first image-forming lens. The image is either formed on a screen that has a hole at its centre through which the illuminating electrons pass (Barnett and Nixon,

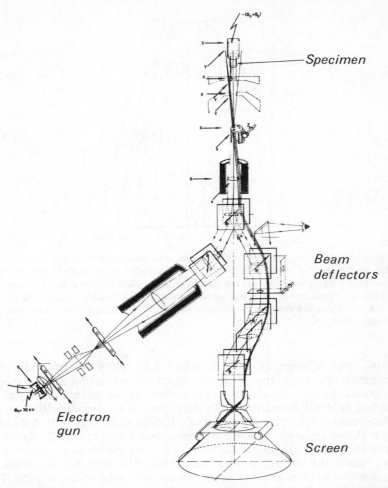

Figure 8.11. Schematic diagram showing the layout of the main components of the reflection microscope designed by Bok. (Courtesy of A. Bok)

1967), or the illuminating and image-forming optic axes are separated by a magnetic prism of the type described in the previous section (Bok, 1970). For the details of these designs and the way in which contrast is formed on the viewing screen the review paper by Bok already referred to should be consulted. Figure 8.11 is a diagram of the instrument devised by Bok, showing its relative complexity.

8.5 SCANNING-TRANSMISSION MICROSCOPE

The standard scanning electron microscope has a resolution that is limited by a combination of the size of the focused electron probe and the spreading of the electrons within the material. This second effect places an ultimate limit on the resolution, so that there is no point in having a focused spot size of much less than 20 nm. However, if the microscope operates in transmission using a thin foil as the specimen, spreading no longer occurs to the same extent and the resolution is potentially higher. As already explained in chapter 2, with a conventional gun any reduction in probe size is obtained at the expense of beam current, and if the time to produce an image is to be kept acceptably short the brightness of the gun has to be increased to take advantage of the potential increase in resolution. A high-resolution scanning microscope using a high-brightness field-emission gun was first constructed by Professor Crewe at Chicago University, and the type of illumination system used in his instrument is described first. This is followed by sections concerned with contrast production and with applications of the instrument. The microscope itself is in essence very simple, consisting of the illumination system above the specimen and a detector or detectors mounted below.

8.5.1 Illumination System

The basis of the illumination system for the high-resolution scanning microscope is a field-emission gun, in which the electrons are obtained from a sharp point, as in the field-emission microscope. Compared with the usual heated filament this gives a considerably higher current density from its surface (typically 10^{10} A m^{-2} compared with 10^5 A m^{-2}), usually from an area small compared with the size of the emitting tip itself. Tungsten is a good material for the tip, since it has low vapour pressure, high melting point, fair electrical and thermal conductivity and is mechanically strong. For the best results the electrode is formed from a wire with a [310] axis since it has been found that this gives the maximum emission along the axis. Since contaminant layers affect field emission adversely, the gun operates under high vacuum and the tip is cleaned initially by heating it briefly to 1900 K. Subsequently whenever the emitted current becomes erratic the tip is heated to 1000 K to restore the original performance. Recent experiments have shown that carbon-fibre tips may also be suitable for use in a field-emission gun. The advantage of this material is that it seems to operate satisfactorily at a pressure of up to 10^{-8} torr, compared with the pressure of 10^{-10} torr necessary in the case of tungsten (English et al., 1973).

The tip is mounted inside a gun composed of a pair of electrodes. These focus the emitted electrons to a small spot, typically ~ 10 nm across, at a distance of 4 cm from the second electrode (Crewe, 1970). Such a gun, by itself, could take the place of the electron-optical column of the standard SEM, giving comparable resolution with a much greater brightness and hence shorter exposure time, but with the considerable disadvantage of requiring a very high vacuum.

In order to obtain very high resolution, it is necessary to reduce the size of this probe to the order of 0.3 nm or so. This is done by a condenser lens that has to be of extremely high quality. Indeed a lens that will demagnify a focused beam by a factor of 50 or so and give a minimum spot size of the order of 0.3 nm will, if used with electrons travelling through it in the reverse direction, enlarge an image by the same factor with a best resolution of about 0.3 nm. In other words the condenser lens has to be comparable in quality with a high-resolution microscope objective lens. Since the performance of such a lens is limited in either application by spherical aberration, it is to be expected that the best resolution of the scanning instrument will be similar to that of a standard high-resolution transmission microscope.

Thus an illumination system, consisting of a high-brightness field-emission gun followed by a single, high-quality condenser lens, is able to give the small probe size needed for high resolution with, at the same time, sufficient total current for the exposure times to be reasonable. As in the normal scanning microscope, two sets of scan coils are located before the condenser lens to enable the beam to be scanned across the sample.

8.5.2 Contrast Mechanisms

This section considers not only the mechanisms of contrast production but also, in outline, how the electrons are collected to form the image, since this is related to the type of contrast that is to be produced. While in principle contrast can be produced by the many processes used in the usual scanning microscope, in practice the largest signal available for detection is the transmitted electron beam. This consists of fast electrons, some of which have passed through the specimen without being scattered, and some of which have been scattered either elastically or inelastically. These electrons can be used, as shown below, to give contrast identical to that obtained from the standard transmission microscope. Alternatively, by modifying the detection system it is possible to enhance this contrast in such a way that it is effectively a new contrast mechanism. In either case, of course, the contrast can be modified by processing the signal electronically before displaying it.

(a) Normal Electron Microscope Contrast
The similarity of the contrast obtained using the scanning microscope to that produced by the normal transmission instrument can be shown with the aid of the Helmholtz reciprocity theorem and figure 8.12. The reciprocity theorem applies to the situation where a ray is able to travel through an optical system from a point P_0 to a point P, and says (Born and Wolf, 1965): 'A point source at P_0 will produce at P the same effect that a point source of equal intensity placed at P will produce at P_0'. Figure 8.10 shows in schematic form the optical arrangement in either the transmission or the scanning transmission microscope: the condenser lens system of the transmission microscope is omitted since it is not relevant to the principles involved.

In the transmission instrument, electrons are produced by the source at P_0 and, after passing through an aperture to limit the total current, fall on the specimen. Those electrons transmitted by the specimen are focused

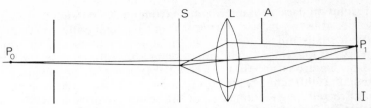

Figure 8.12. Schematic representation of the layout of the essential components of a microscope/scanning microscope: the electrons travel left to right in a
conventional microscope and in the reverse direction in the scanning instrument

by the high-quality objective lens to form a diffraction pattern in the back
focal plane of the lens where the objective aperture A is located, and a
magnified image in the plane I. P_1 is a typical point in the image. In the
scanning instrument, the beam is scanned over the specimen by effectively moving the source of electrons. In this case P_1 represents one of the
positions of the source during the scan, and the aperture A is now the
aperture of the condenser lens, which projects a demagnified image of
the source onto the specimen S. A detector is placed at P_0. By following
the diagram from left to right or right to left it can represent either a
transmission or scanning transmission microscope. In order to calculate
the contrast for the transmission instrument we have to find the intensity
at a point such as P_1 for a given source intensity at P_0, whereas in the
equivalent calculation for the scanning transmission case we have to
calculate the intensity at P_0 when the source of illumination is at P_1. For
the simple arrangement considered here, according to the reciprocity
theorem the image intensity is in each case the same and therefore the
image contrast is identical. In practice of course, the sources do not have
the same intensity, but the ratio of the two image intensities is constant,
so that the contrast is the same using either instrument.

From this it can be seen that the conditions under which the two
microscopes will give the same contrast are (a) the angular divergence of
the illumination in the transmission microscope must be the same as the
angular range of acceptance of the scanning microscope detector, and (b)
the microscope objective aperture must be identical in size and in location
in the back focal plane of the lens to the size and position of the scanning
microscope condenser aperture.

Under these conditions it is possible to observe all the effects considered in chapter 5. For example, by defocusing the scanning condenser
lens a little (corresponding to going out of focus in the transmission
microscope) phase contrast effects, such as Fresnel fringes, are visible in
the image, with the added advantage that the image is produced in a form
suitable for computer processing, should it be desired to extract additional information by such means. Again, dark-field images of crystalline
samples are obtained if the detector aperture is displaced to a position
such that, if it were the transmission microscope source aperture, the
diffraction spot formed with it in that position would lie at the centre of
the aperture A: this is equivalent to using tilted illumination to get

high-resolution dark-field micrographs from a transmission instrument. It is also possible to observe diffraction and Kikuchi patterns. In order to observe a single diffraction spot in the normal microscope, it would be necessary (a) to reduce the illumination aperture to improve the collimation of the electrons, and (b) use a small objective aperture displaced so as to admit the diffracted beam in question. With the same geometry the same diffracted beam is detected in the scanning instrument. By moving the aperture A in its plane, each of the possible diffracted beams is detected in turn: in practice, the scan coils would be operated in the same way as is used to obtain a channelling pattern in the SEM, and so rocking the beam about a particular point on the specimen, rather than moving the aperture.

The treatment of scanning microscope contrast as outlined so far does not take any account of the inelastic scattering of the electrons. Pogany and Turner (1968) have shown that, subject to the (good) approximation of neglecting the change in wavelength of the inelastically scattered electrons, the reciprocity theorem holds in the presence of inelastic scattering. Thus, inelastic scattering does not in principle give rise to any difference between the images formed by the two types of microscope. However, in practice there is a difference, because in the normal microscope the image-forming lens is located after the specimen, while in the scanning instrument it comes before. Whereas the detector in the scanning case accepts all the electrons irrespective of their energy, in the normal electron microscope the objective lens, which suffers from chromatic aberration, is not able to focus all the electrons into a sharp image simultaneously. Thus the resolution of a thick specimen is limited by chromatic aberration in the case of the transmission microscope but for the scanning microscope it is limited by the much smaller effect of the spreading of the beam as it travels through the specimen (figure 8.13). This implies that, while the resolution of the two instruments should be similar for very thin specimens, the scanning instrument should be increasingly superior as the sample thickness increases.

(b) Additional Contrast Mechanisms

By using a detector that is more complicated than the simple detector implied in the description given in the previous section, or by processing the image after collection, it is possible to produce contrast which the ordinary microscope cannot give. If an energy analyser is mounted in front of the detector and only electrons within a narrow energy band are allowed through, the resulting image will be similar to that obtained using an energy-selecting microscope. Comparing these two instruments, the scanning microscope has the advantage of fewer electron-optical components but the disadvantage of requiring an extremely good vacuum. If the signals corresponding to two characteristic losses can be separated and collected simultaneously, it should be possible to present images, one due to electrons scattered in one phase of a multiphase system, and one to electrons from another, side by side, a refinement not possible with the energy-selecting instrument.

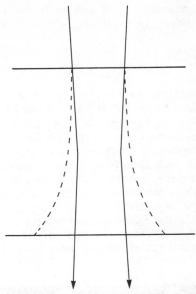

Figure 8.13. Schematic representation of the spreading of the electron beam as it travels through the specimen in the scanning microscope. The full line shows the region through which the unscattered electrons travel: the scattered electrons travel in paths which lie within the dashed curves

As noted in chapter 5, the absorption contrast from very thin specimens is extremely weak, and while it is possible to increase this with the scanning microscope either by using characteristic loss electrons (only a few from a thin specimen) or by adjusting the effective intensity zero electronically (and increasing the image noise), Crewe (1970) has devised a simple method of increasing this contrast. It involves using a detector that does not accept any of the unscattered electrons: it collects only those electrons that have been deflected away from the optic axis, and is therefore essentially a dark-field technique. These electrons are analysed very crudely into those that have not lost any energy and those that have, and the ratio of these two signals taken by dividing one by the other electronically. This ratio is approximately proportional to Z, and can be displayed on the monitor screen to give an image whose contrast is rather better than the normal image. The contrast is increased because the signal is proportional to Z, so that areas differing in atomic number by, say, 5 in 30, will be appreciably different in contrast even without any electronic processing of the signal. In addition, this contrast can with confidence be interpreted as being due to difference in atomic number, since the ratio is not affected by fluctuations in the incident electron intensity at the specimen, or (for thin samples) by variations in the specimen thickness. More sophisticated applications of these ideas, involving the use of more detectors at different angles to the optic axis, or analysing the energies of the electrons more accurately, can obviously be used to enhance the overall contrast or show up particular elements or combinations of elements.

8.5.3 Applications

Since at the time of writing, this microscope is still in the development stage as a commercial instrument, it has not yet been applied to many research problems. It is, however, possible to see, on the basis of its particular advantages, in what areas it is likely to be especially useful. Its chief advantages compared with the standard transmission microscope are similar resolution and potentially higher contrast when used as a high-resolution microscope, and the ability to give adequate resolution of structures such as dislocations in much thicker specimens. The possibility of resolving the shape of individual molecules, and possibly observing individual atoms within the molecule, at very good contrast has already been demonstrated (figure 8.14a). Undoubtedly this facility will be very useful, both to biological scientists who wish to study organic matter at

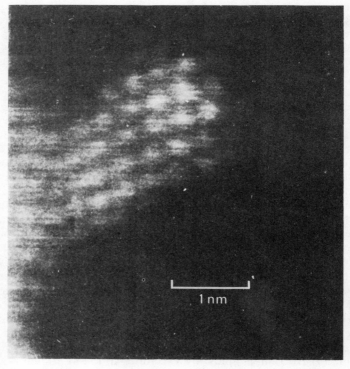

(a)

Figure 8.14. (a) Image of a thin UO_2 crystal, showing a spacing of $\sim$3.5 Å between atom spots. (Courtesy of A. V. Crewe)

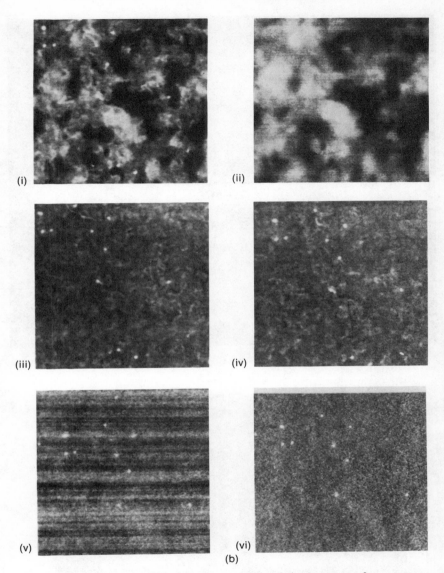

Figure 8.14 (contd.). (b) Set of micrographs of single atoms of mercury on a 2 nm thick carbon film, showing the improvement in image quality possible when using special techniques to produce the contrast. (i) 'Elastic' image; (ii) 'inelastic' image (some bright spots just visible in the elastic image); (iii) ratio of elastic to inelastic; (iv) elastic–Kx inelastic intensity (K = elastic/inelastic for the carbon film); (v) the ratio of outer annular detector current to inner annular detector current for a double annular collector system. The scanning stripes are due to leakage currents in the detectors; (vi) is the image (v) differentiated electronically. In each case the image width corresponds to 20 nm. (Courtesy of A. V. Crewe)

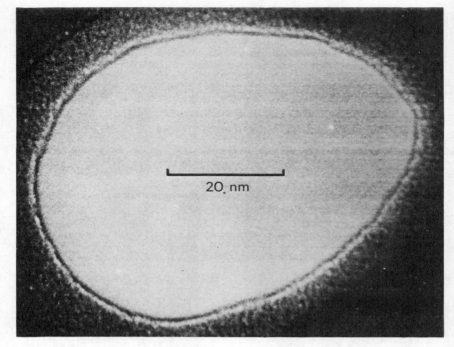

(c)

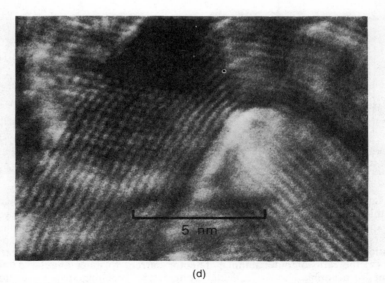

(d)

Figure 8.14 (contd.). (c) A Fresnel fringe around a hole in a carbon film; (d) an image of 0.34 nm lattice plane fringes in graphitised carbon. (I. F. L. Ray, courtesy of A.E.I. Ltd)

this level, and to physicists and materials scientists who wish to study processes such as nucleation and growth of islands on substrates. Indeed, for this last type of application the need to operate the microscope at high vacuum will be an advantage, rather than a drawback.

So far as employing the improved resolution from very thick specimens is concerned, this will be of especial use to those who wish to observe the three-dimensional arrangement of microstructures of both biological and materials-type specimens. Again the ability to improve the contrast or obtain contrast by using inelastically scattered electrons in various ways will enable the microscope to extract information from the specimen inaccessible to the standard transmission instrument. Some other examples of micrographs obtained with this instrument are shown in figures 8.14b–d.

The main disadvantages of the microscope are the need to work at ultra-high vacuum because of the gun design, and the cost, which seems likely to be rather greater than that of a standard instrument. For many applications these will be outweighed by the advantages, and these microscopes are likely to fairly widespread in the future.

In particular there seem to be some advantages in combining the instrument with a conventional transmission microscope so that the type of image produced may be changed rapidly from one sort to the other. Commercial instruments of this design are becoming available at the time of writing.

8.6 GUIDE TO FURTHER READING

There are a number of books and articles in which instruments of the type considered here are described. These include Amelinckx *et al., Modern Diffraction and Imaging Techniques in Material Science* and Valdrè (ed.) *Electron Microscopy in Material Science*, which have both been referred to on specific matters in the text; also various editions of *Advances in Electronics and Electron Optics* and *Advances in Optical and Electron Microscopy*.

Many of these instruments are still in the development stage and improvements are being made all the time: these are reported either in the research journals, such as *J. Phys. E.* (formerly *J. scient. Instrum.*) and *Rev. scient. Instrum.*, or else at conferences, such as the *International Conferences on Electron Microscopy*, and the *European Regional Conferences on Electron Microscopy*, (these are both held every four years, with a two-year gap between one type and the other).

New applications for existing instruments tend to be reported in the journals and conferences concerned with materials studies, as well as the journals and conferences mentioned above.

REFERENCES

Amelinckx, S., Gevers, R., Remaut, G., and van Landuyt, J., (eds) (1970), *Modern Diffraction and Imaging Techniques in Material Science*, North-Holland, Amsterdam

Barnett, M. E. and Nixon, W., (1967), *J. scient. Instrum.*, **44**, 893

Bok, A., (1970), *Modern Diffraction and Imaging Techniques in Material Science*, (eds Amelinckx *et al.*), North-Holland, Amsterdam, 655

Born, M. and Wolf, E., (1965), *Principles of Optics* (3rd edn), Pergamon, Oxford, 381

Castaing, R., (1971), *Electron Microscopy in Material Science* (ed. U. Valdrè), Academic Press, New York, 102

Castaing, R., and Henry, L., (1962), *C.r. hebd. Séanc. Acad. Sci., Paris*, **255**, 76

Crewe, A., (1970), *Q. Rev. Biophys.*, **3**, 137

Cundy, S. L., Metherall, A. J. F., and Whelan, M. J., (1966), *J. scient. Instrum.*, **43**, 712

Curtis, G. H. and Silcox, J., (1971), *Rev. scient. Instrum.*, **42**, 630

English, T. H., Lea, C. and Lilburne, M. T., (1973), *Scanning Electron Microscopy, Systems and Applications*, Institute of Physics, Bristol

Gadzuk, J. W., and Plummer, E. W., (1973), *Rev. mod. Phys.*, **45**, 487

Gomer, R., (1961), *Field Emission and Field Ionisation*, Harvard University Press

Good, R. H. and Müller, E. W., (1956), *Handb. Phys.*, **21**, 176

Graber, R., Gribi, M. and Wegmann, L., (1968), *Proceedings of the 4th European Regional Conference on Electron Microscopy*; Tipografia Poliglotta Vaticana, Rome, 111

Kinsman, K. R. and Aaronson, H. I., (1972), *Electron Microscopy and Structure of Materials*, (ed. G. Thomas), University of California Press, 259

Melmed, A. J., (1965), *J. appl. Phys.*, **36**, 3585

Melmed, A. J., (1967), *J. appl. Phys.*, **38**, 1885

Müller, E. W., (1949), *Z. Phys.*, **126**, 642

Müller, E. W., (1970), *Modern Diffraction and Imaging Techniques in Material Science*, (eds Amelinckx *et al.*), North-Holland, Amsterdam, 683

Pogany, A. P. and Turner, P. S., (1968), *Acta crystallogr.*, **A24**, 103

Silcox, J., and Vincent, R., (1972), *Electron Microscopy and Structure of Materials*, (ed. G. Thomas), University of California Press, 259

Swanson, L. W., and Bell, A. E., (1973), *Adv. Electronics Electron Phys.*, **32**, 193

Valdrè, U., and Zichichi, A., (eds) (1971), *Electron Microscopy in Material Science*, Academic Press, New York

Wegmann, L., (1972a), *Electron Microscopy and Structure of Materials* (ed. G. Thomas), University of California Press, 246; (1972b), *J. Microscopy*, **96**, 1

Zaminer, Ch., (1968), *Proceedings of the European Regional Conference on Electron Microscopy*, Tipografia Poliglotta Vaticana, Rome, 123

Index